Fundamentals of GPS Receivers

Dan Doberstein

Fundamentals of GPS Receivers

A Hardware Approach

 Springer

Dan Doberstein
DKD Instruments
750 Amber Way
Nipomo, CA, USA
dand@dkdinst.com

Please note that additional material for this book can be downloaded from
http://extras.springer.com

ISBN 978-1-4614-0408-8   e-ISBN 978-1-4614-0409-5
DOI 10.1007/978-1-4614-0409-5
Springer New York Dordrecht Heidelberg London

Library of Congress Control Number: 2011938456

Printed on acid-free paper

Springer is part of Springer Science+Business Media (www.springer.com)_

Preface

If one examines the current literature on GPS receiver design, most of it is quite a bit above the level of the novice. It is taken for granted that the reader is already at a fairly high level of understanding and proceeds from there. This text will be an attempt to take the reader through the concepts and circuits needed to be able to understand how a GPS receiver works from the antenna to the solution of user position.

To write such a text is not trivial. It is easy to get distracted in the GPS receiver. Many papers and articles deal with the minutiae of extracting the last little bit of accuracy from the system. That is not the goal of this text. The primary goal of this text is to understand a GPS receiver that solves for the "first-order" user position. What is meant mean by "first-order solution"? The best way to answer that question is another question and that is, "What do we have to do to build the minimum GPS receiver system to give user Position accurate to approximately 300 m?" The reader should know that as desired accuracy of position or time solutions increases so does the complexity of the receiver. In pursuing the 300-m goal, the reader will gain an understanding of the core principles present in all GPS receivers. It is hoped that the reader will then be able to proceed from there to understand the later techniques presented that achieve accuracy above this level.

A major problem in writing this text is the assumed background of the reader. It is not possible inside this text to start at receiver fundamentals and work from there. An assumed background level is needed. The basic background of the reader should include an understanding of analog narrow-band radio receivers, basic digital circuits, algebra, trig and concepts from calculus. The solution of the equations for user position is the most challenging in terms of the math needed. Linear algebra and calculus are used.

Regardless, it is not the intent of this text to smother the reader in math, equations, and the like. A more practical approach will be pursued. An attempt will be made to describe the concepts and phenomenon with as little math as possible. It is impossible to write such a text without equations so where appropriate they will be used.

GPS receivers must solve two fundamental problems: First is the receiver itself, which gets the raw range and Doppler to each SV (the observables), the second is the manipulation and computations done on that data to calculate the user position. These two problems are intertwined in such a fashion which makes complete separation impossible. It would seem natural to start at the receiver antenna and work backward into the receiver. But this approach does not lay the needed foundation of understanding of basic principles at work in the GPS. Instead, the text will be split into three parts.

Part I will introduce the reader to fundamental process behind all GPS receivers. Simplified models will be used wherever possible. The details of the GPS signal and its data stream will be explored. With this knowledge, the solution of users position will be presented without getting into the details of the receiver hardware. Therefore, understanding the Part I of this text does not require the reader to have intimate knowledge of radio receiver methods.

Part II explores the details of the receiver. The reader will need to understand radio principles very well to completely follow the discussions presented. This text will develop receiver concepts using a hybrid design. Although most commercial (if not all) GPS receivers today use DSP methods, it is the author's view that these techniques are difficult to learn the fundamentals from. The approach pursued in this text is just easier to understand. Digital methods will be used and their analog counter part, if any, will be discussed.

In Part III, more advanced receivers and topics are covered. In Chap. 8, we will examine GPS time receivers, time and frequency measurements using GPS receivers and simple time transfer. In Chap. 9, the Zarlink GPS receiver chip set is discussed as introduction to more modern receiver using DSP methods. In Chaps. 10 and 11, the most advanced material is presented with the majority of the material focused on Carrier Phase Methods. Chapter 11 discusses the Turbo Rogue Receive, one of the most accurate GPS receivers ever made. In Chapter 12 the new GPS signal L2C is detailed along with receiver methods for L2C signal. Chapter 12 is contributed by Danilo Llanes.

As a final comment, many readers may come to this subject with the idea that GPS is only about the physical position of the user and satellites. As one learns more about GPS it becomes apparent what GPS is really about is *Time and Movement*. The GPS receiver uses observations of Time (Clocks) to *Measure* movement. The result is that the electronic clock signals as received, inside the receiver, will also be found to be moving in Time in direct relation to the physical movement of the receiver/satellite system.

Nipomo, CA, USA   Dan Doberstein

Contents

Part III

Part I

Chapter 1
Fundamental Concepts of Distance Measurement Using Synchronized Clocks

1.1 The Fundamental Process of Measuring Distance

Before we start on our journey we need to form a mental model of the fundamental processes that must be done to get to our goal, the Users Position on our earth. There are four distinct processes at work in the GPS receiver.

- The first process is reception of the signal itself.
- The second process examines this signal, acquires lock on the signal, and retrieves the satellite data.
- The third process measures the distance from the user to each satellite the user receiver tracks.
- The fourth process is the calculation of user's position using information from above processes.

As stated in the introduction we will not follow the above processes in sequential order. Instead Part I of this text will cover topics associated primarily with the third and fourth processes. Part II of the text will concentrate on first and second processes.

As we shall shortly see the fundamental problem in GPS is not measuring distance but instead one of measuring *time*. In fact the deeper one "digs" into GPS the more apparent it becomes that solving the time issues is really the whole purpose and function of the system.

1.2 Comments on the Use of Models and the ICD-200 Document

Throughout this text we will make liberal use of various models to help explain and understand GPS. The reader should be aware that some models are distantly related to actual function while others are very close approximations to the actual system or

D. Doberstein, *Fundamentals of GPS Receivers: A Hardware Approach*,
DOI 10.1007/978-1-4614-0409-5_1, © Springer Science+Business Media, LLC 2012

sub-system being discussed. Regardless when the term "Model" is used, we will be making some approximation or simplification of GPS or related phenomenon. If the reader wants a precise definition of GPS, the ICD-200 document is the place to go. Its "chock-full" of information arranged in such a fashion that it is difficult to "see the forest for the trees." The diagrams are precise and complex. The few models used are quite intricate. The author highly recommends reading this reference, as it is the ultimate authority on the specifications of GPS.

1.3   Distance Measurement by Time of Arrival Measurement

We can start our discussion of distance measurement with a very simple example. Two people playing catch have the ability to measure the distance between themselves if they know the speed of the ball and the time it takes to pass from one to the other. We can equip the pitcher with a special watch that stamps the time on the ball when it is released from the hand. At the catcher another watch stamps the ball with time received as it strikes the glove. On inspecting the ball we can see the time sent and the time received. If the two clocks are synchronized to read the same time beforehand, the difference in the two times stamped on the ball is the transition time from pitcher to catcher. If we know the speed of the pitch we can compute the distance between the two players from distance = speed × time.

It may seem simplistic but this example illustrates the fundamentals behind GPS position determination. In GPS the "ball" is replaced by an encoded radio wave. The encoding contains the time sent or more precisely the GPS clock information, which is delayed in time due to the path delay from the satellite to the Earth. The speed of the radio wave is the speed of light. In the GPS, satellites are each pitching a ball to the user's receiver. The user's receiver can use the time sent/received information and the speed of light to compute the distance to each satellite. Additional encoding on the signals from the satellites tell the user's receiver the position of each of the satellites in the sky in x, y, z. Then equations are solved to determine the user's position from all this information.

Sounds simple. But in practice this is a very complex operation. And at the heart of all this complexity lies one of the fundamental issues with this approach to position measurement: Clock Synchronization. In the simple pitcher/catcher example above we assumed the clock at the pitcher and the catcher were synchronized to read the same time, that is, they read the same time if examined at the same instant. Clock Synchronization is not easy to achieve for GPS. It is difficult because the GPS signal is traveling at the speed of light so that even an extremely small misalignments or errors in the synchronization of the clocks used (one in each satellite) will translate to large errors of computed distance and hence position. So methods must be devised to synchronize all the clocks involved very precisely, both at the user and at the satellite.

Later in this chapter we will form some simple linear models that will allow a simplified exploration of some of the complexities of this problem. The math involved with a linear model is far easier than that of the multi-satellite system in 3-D and almost all the phenomenon associated with the 3-D system can be worked out. Before we move on to those linear models lets take a closer look at clock synchronization and how we use synchronized clocks to measure the distance from a user at the earth to a GPS satellite.

1.4   The Physical Process of Clock Synchronization

It would seem that synchronizing two clocks should be easy. For everyday measurements like agreeing to meet another person at particular time synchronization is no problem. The error between the two clocks will be insignificant for this purpose. But in GPS we wish to measure how far a light beam travels in a given amount of time. This calls for extremely precise clock synchronization due to the high speed of light. But just exactly what is meant by "synchronization" of two clocks? This may seem like a silly question until one thinks about what actually occurs when we "set" our watch to the clock on the wall. When we set (or synchronize) our wristwatch to a wall clock we are inadvertently making a very small error. The time we set on our wristwatch is actually the wall clock time as it was in the past. This is because the light that comes from the wall clock takes a finite amount of time to reach our eyeball. If the two clocks are moving or in a gravitational field it gets even more complicated due to relativistic effects. Fortunately the relativistic effects involved in GPS turn out to be small and for most user purposes these effects can be ignored.

So, in order to truly synchronize two clocks that have any distance between them you must take into account the time it takes for the information from the clock you are synchronizing with to get to you. In short, you must know the distance between two clocks to perform truly accurate clock synchronization. Therein lies the heart of the problem, we wish to measure distance using two clocks a Radio Frequency beam. But we must know this very distance to precisely set our clock! We will see later how this problem is overcome.

It is hoped that the reader can now see that synchronizing two clocks precisely is indeed non-trivial. The whole process of defining "clock synchronization" when we must consider the finite speed of light is quite troubling. There is another way. If we imagine that we can "see" the clocks we wish to synchronize with light that has infinite speed then we can more easily define synchronization. With our infinite speed light beam the distance between the clocks is no longer a concern and we can set our clocks to read the same time no matter how far apart they are. This is what is truly meant by clock synchronization as used in GPS. In an *ideal* GPS satellite constellation looking at each satellite from the ground you would see the same time on every satellite clock as long as "infinite" speed light is used to view the clocks.

1.5   Magic Binoculars

As we have alluded to above each GPS satellite has its own clock. We need to think about what time a person on the earth would "see" on the GPS satellite by using a special pair of binoculars. These binoculars are magical in that the left eye sees the satellite clock with infinite speed light. The right eye sees the satellite clock with light that travels the normal "fixed" or finite speed. So what would we see in our magic binoculars? The left eye sees the time on the satellite exactly as it as it happens as no delay exists due to infinite speed of light for this eye. The right eye would see the clock displaying a time in the *past* as this light is "normal" light that travels at finite speed. The difference in time between the time seen by the left eye and the time seen by the right eye is the path delay from the person on the earth to satellite. By multiplying this delta time by the known speed of "normal" light we can compute the distance from the person to satellite.

We can eliminate our need for the special "infinite" speed side of the binocular if we have a synchronized clock (reference clock) on the ground next to us when we look up at the satellite clock with our monocular or telescope. This reference clock must be displaying the time we would see through the "infinite speed" side of our magic binoculars. By looking simultaneously at the satellite clock through the telescope with one eye and the other eye looking at our synchronized clock, we can do the same computation we just did with our "magic binoculars."

Of course a human is not fast enough to do these tasks without special machines that allow quick and precise capture of the clock times involved. The purpose of these thought experiments is to convey the concepts behind the process. In particular, the concept and use of synchronized clocks to measure distance by measuring path delay experienced by light and exactly what is meant by the term "synchronizing."

As we can see already the GPS receiver is really about using clocks to measure time differences. Once these time differences are known for four Satellite Vehicles (SV's) we can compute the corresponding distances which will lead to a computed user position. The real task of the GPS receiver is clock synchronization. The receiver must sync his own reference clock and also replicas of SV clocks.

In practice, the user's receiver has at least two clocks. One clock will be used to display GPS time and the other clock is a replica of the clock on board the SV. This would be called a single channel receiver, as only one SV replica clock is present. This replica clock is running at very nearly the correct rate but its displayed time will not be correct until the user receiver "synchronizes" it. Decoding the information that is sent to the user on a radio frequency beam allows the replica clock to be "synced." The radio beam is encoded with SV clock timing. Once we properly decode the timing signals, the user receiver can "set" its replica clock such that the displayed time is very nearly that of the SV but *delayed* by the time it takes for the radio beam to traverse the distance from SV to the user.

1.6   A Simple Light Pulse Transmitter and Receiver to Measure Distance

Now let us look at the problem of determining the distance of a car from a known point using light pulses. We will assume the road is perfectly straight and level. We wish to determine the distances from our car to the beginning of the road at point A and from end of the road at point B, see Fig. 1.1. We desire the accuracy of the position measurement to be about 300 m. The total length of the road is 6,000 km. At point A and in your car are clocks that count from zero to 20,000 μs (or 20 ms) and then start over again. One microsecond resolution is chosen as light travels 300 m in 1 μs, which is our desired resolution in distance. The choice of road length is determined by the fact that light travels 6,000 km in 20 ms. A longer stretch of road would introduce an *ambiguity* in the time measurement. Shortly we will expand our clock for longer *unambiguous* distance measurements.

We will assume perfect synchronism between the two clocks, in other words they read zero at EXACTLY the same instant. Every time a clock counts to 20,000 it rolls over to zero and starts over again. At the point in time where the clock at point A rolls over (time zero) a light pulse is emitted. As you travel the highway in your car you would see flashes of light from the rear every 20 ms. The moment a pulse of light from point A "hit's" the receiver in our car the time indicated on the cars clock is photographed by camera A. The photograph of the car's clock from this moment indicates the *Time of Arrival* (TOA) of the light pulses from point A. Since we have assumed perfect synchronism of the two clocks and we know that the pulse was sent from point A at time zero we can compute the time of travel from

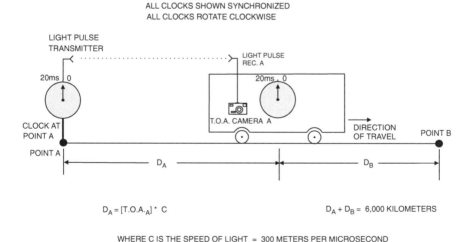

Fig. 1.1 Car on roadway example with light pulse transmitter and receiver

Point A to our car by simply reading the time from the photograph. Also we know that the pulse traveled at the speed of light. So we can easily compute the distance to point A or D_a by;

$$D_a = \text{(time on car clock when pulse A arrived in integer number of microseconds)}$$
$$\times \text{ speed of light} \qquad (1.1)$$

By knowing that the pulse left point A at precisely time zero on all clocks and that the two clocks are perfectly synchronized we can compute the distance to point A as we travel down the roadway.

1.7 Problems with the Simple Light Pulse Transmitter/Receiver System

Our simple model of Fig. 1.1 has a few shortcomings. The first issue is range of the clocks used. They can only read from 0 to 20 ms. This limits the maximum unambiguous range to 6,000 km. Any distance past this, we would have to sort out which pulse left at what time. That is not a problem we wish to solve or discuss here. Because the maximum distance from the user receiver on the ground to an orbiting GPS satellite is ~25,000 km, we desire the clock to cover more time to avoid the ambiguity issue. Additionally a 1 μs resolution limits our position accuracy to approximately ±300 m. If we used a 0.1 μs or better resolution, we resolve distances down to about ±30 m.

The second problem is that we assumed that a pulse left at exactly "time = 0" from point A. This made it easy to compute the time of travel of the emitted pulse, it was just the TOA as recorded on the car's clock. We seek a new system that will allow us to determine when the pulse was sent without the requirement that it leave the transmitter at time zero.

In summary, what we need is a new model that has an expanded clock range to resolve the limited time range issue and a finer clock resolution for better position resolution. In addition, we need to devise a way to tell the receiver when a pulse (or "timing edge") was sent so we can send them at other times besides $t = 0$.

1.8 A New Clock Model

Figure 1.2 shows our new clock. This clock is a very close model to that actually used by the GPS. As we progress in the text we will add refinements and comments about this clock, as we need them. For now and the majority of this text this clock will serve all our needs.

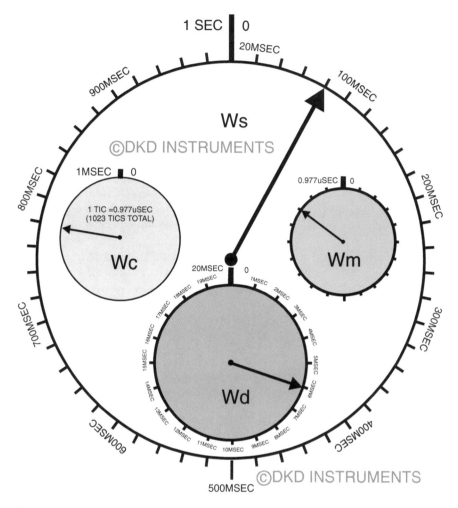

Fig. 1.2 Precision clock for measuring up to 1 s intervals with sub micro-second accuracy

The clock is composed of a main "Dial", W_s and three sub-Dials, W_d, W_c, and W_m. It works like an old fashion mechanical clock in that each hand of all four Dials moves in a series of "tics." The main Dial tic size is 20 ms. It covers 1 s total so there are 50 tics in this Dial. The next Dial down in tic size is the 0–20 ms Dial. It has 20 tics so each tic is 1 ms. The next smaller Dial in tic size is the 0–1 ms Dial. It has 1,023 tics with each tic being ~0.977 μs of time. The number of tics is so large on this Dial the figure cannot show them. This funny choice of the tic size (and number of) for this Dial will be explained in the following chapters. The choice of the dial names will become clearer as we progress.

The final Dial measures the smallest time interval of our clock covering just 0–0.977 μs. The tic size of this Dial depends on the exact hardware implementation of the receiver. The resolution of this Dial also depends on if we are talking about a SV clock or a terrestrial-based version of it as implemented in the user receiver.

For user receiver to achieve a distance resolution of ~30 m, we would need a tic size of ~0.1 μs or about ten tics as discussed above. Many modern "digital" GPS receivers may have a tic size here (or equivalent) as small as ~3 ns. In the hardware section of this text we will explore, in detail, a method to produce a tic size of 48.85 ns which results in this Dial having 20 tics. As we move ahead in our discussion we will address the resolution of the 0.977 μs Dial as implemented in the SV.

Our clock works just like the stopwatch in that one full revolution of the 0 top 0.977 μs Dial produces a one-tic movement on the 0–1 ms Dial. Likewise, a full revolution of the 1 ms Dial produces one tic movement on the 20 ms Dial. Lastly, a full revolution of the 0–20 ms Dial produces a one-tic movement on the 1–s Dial.

We now have a clock that has the range and resolution we need to accurately measure the signals used in GPS. The clock is a "digital" clock in the sense that all movement is in "tics." If we take a snapshot of the clock and wish to compute the time it presents as a single number we can do the following calculation:

Time on Clock = (1 s Dial Tics × 20 ms) + (20 ms Dial Tics × 1 ms) + (1 ms Dial Tics × 0.977 ns) + (0.977 Dial Tics × 48.85 ns)

IMPORTANT!: *In our model the W_m dial is limited to ~48 ns resolution. The actual GPS system clock has MUCH finer resolution than this. In this text, unless otherwise specified, we will work with this resolution.*

1.9   A "Time Transfer" Linear Model

By improving our clock we have addressed the issues of range and resolution of our time measurements. We still have the "time sent" issue to resolve. One way to address this issue is to *send* the clock information on a light or radio beam to the receiver that needs it.

In our light pulse example above we used light pulses sent from point A and B to measure the distance of our car. GPS *does not* use light. Nor is GPS a *pulsed* system, strictly speaking. To make a more faithful car/roadway model of the GPS system, we can encode the light beams such that the receiver in the car can re-construct a replica of the clock from point "A." In other words, we send the clock information present on clock A to the receiver on the car. The receiver decodes the information and forms a replica of the remote clock "inside" the receiver. The question of how such encoding can be done will have to wait. For now on the reader is asked to assume that this can be done.

Figure 1.3 shows our new linear model using our new clock. The car now has two clock displays, the receiver clock and a replica of clock "A" that is

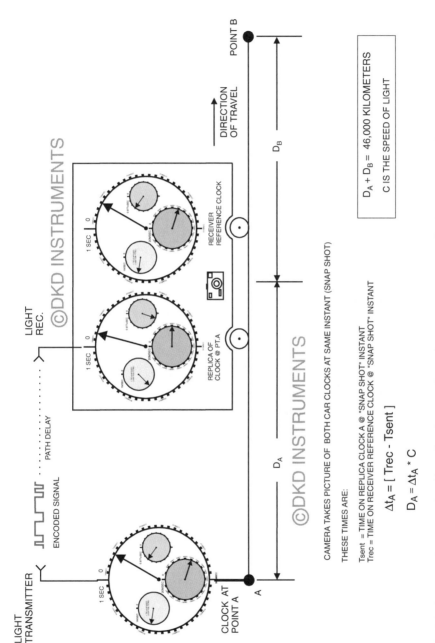

Fig. 1.3 Car on roadway example with replica clock, receiver reference clock is synchronized with clock at point A

"reconstructed" from the received light signals from point "A." We will assume that the receiver car clock is synchronized to the clock at A. The receiver's replica of clock "A" is offset from the time at "A" by the path delay from point A to the car. This delay is caused by the finite speed of light.

The replica clock on the car shows the time from clock A as it was in the "past." *The observer in the car is seeing the "time" on the replica clock from the point in* *time when the encoded light signal left point A. In other words, the receiver's* *replica of clock "A" is displaying the "time sent" information.*

As we travel along in the car we can take a photo of the two clocks at any time we wish to record their "state." The point in time that the "snapshot" occurs does not have to be synchronous with any clock time and can happen at any instant we choose. *The photo must show both clocks on the car to be a valid measurement of* *the delay from A to the car at that instant of time.* By examining the resulting photos, we can determine our send and receive times. From these times we can easily compute the car's position on the roadway by doing the calculations indicated in Fig. 1.3.

Another way to understand what is happening in this model is to think about what would happen if the car was 1 m away from point A. It takes about 3 ns for light to travel 1 m. In this case a photo of the two receiver clocks would show replica A clock indicating a time 3 ns *before* the time indicated on the receiver's synchronized clock (commonly called the receiver reference clock). In other words an extremely small difference between Clock A and the receiver's replica A clock would be present. In order to detect this, extremely small time difference of the 0.977 ms Dial would need a tic size less than 3 ns!

If we move the receiver 300 m away from point "A," the receiver's replica clock will show a time ~1 μs in the past. For the receiver's replica clock A to read *exactly* the time indicated on Clock A, there would have be zero distance between the receiver's replica Clock A and Clock A.

1.10   Clock Synchronization

The comments above point out the difficulty in synchronizing clocks in general. Fortunately for our purposes here we only wish to resolve distance to about 30 m. This is doable with a clock precision and synchronization of about 0.1 μs. In addition, relativistic effects can be ignored, as they result in distance errors less than 1 m in GPS.

To summarize, we need only achieve a synchronization of approximately ±0.1 μs between Clocks A, B, and the Receiver clock to be able to resolve distance to approximately ±30 m.

1.11   Time Transfer Linear Model with Receiver Clock Not Synchronized to Clocks A and B

In our previous two examples we have assumed the receiver reference clock and the clocks at point A to be in perfect synchronism. This key assumption was made so that we could measure the "time received." For a practical GPS receiver this creates a problem. Due to the accuracys needed having a clock in a small low-cost receiver that could be "set" and let free run is not possible. It is only feasible to implement the "free run" approach with large, extremely expensive atomic clocks. So a way must be devised to synchronize the receiver's clock using measurements the receiver can make on its own.

This can be done in our linear model by adding another clock at point B of the roadway, see Fig. 1.4. This clock is assumed to be in perfect synchronism with the clock at point A. We can think of our new system as a linear version of the GPS with just two satellites, one at A and another at B. Our receiver must also be enhanced with another replica clock for the added clock at point B. The receiver now has three clocks, the reference clock and two replica clocks as received from A and B.

The camera will now record the state of the two replica clocks and the receiver's reference clock at the "snapshot" instant. The information in the photo now contains time received and time sent information for the signals from A and B.

We will assume that the error on the receiver's reference clock to be completely random. In other words the four dials can be in any configuration when it is turned on. The receiver needs to set all the dials so that it is in synchronism with clocks at A and B.

Before we proceed with the solution to the EXACT error on the receiver's reference clock, we need to add a bit more information about our car's position on the 46,000 km track. We will assume that we know our position to be somewhere between 20,000 and 25,000 km from point A. In case the reader wonders where these numbers come from it is the approximate minimum/maximum distance from the earth's surface to the GPS SV. With this knowledge we can predict the delay from point A or B to be between 60 and 80 ms.

The receiver can now examine the two replica clocks and set the second hand Dial of the reference clock to be synchronous with the second hands at clocks A and B by adding 60 ms. In other words, we can "sync" our receiver reference clock to within 20 ms directly using our assumed prior knowledge of the car's position on the track. This is what is shown in Fig. 1.4. The second hand of the receiver's reference clock is in the same position as clocks at A and B. The other Dials are still not set and are out of position with respect to A/B clocks.

We can use the redundant information about our position represented by the clock at point B to solve for the remaining error in our receiver reference clock. Following the equations of Fig. 1.4, we can introduce an error term for the receiver reference clock, Tbias. After combining the equations shown in Fig. 1.4, an expression for Tbias can be derived that contains all known or measured terms. In other words, we can mathematically solve for the clock error of the receivers

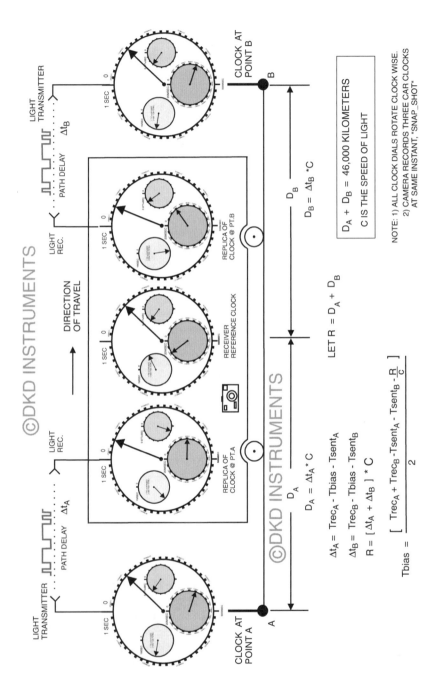

Fig. 1.4 Car on roadway example with replica clock, receiver clock has error (Tbias) with respect to clocks at A and B. Clocks A and B synchronized

reference clock. With this knowledge we can then solve for our position on the roadway. In this fashion car position and receiver clock error can be determined simultaneously. Figure 1.5 shows the receiver with its clock corrected to read the same as those at points A and B. The choice of the sign of Tbias in the equations of Fig. 1.4 is arbitrary. The sign chosen reflects that followed by the sign convention used by GPS. We will continue to follow the sign conventions as established by GPS.

We could have left the clock with *all* its errors intact and still solved for position and clock error. In fact, it is always needed to "physically" correct the receiver reference clock. We can just use the derived error term and add or subtract away this amount of time for the time indicated on the receiver reference clock to know the time at A or B.

It is hoped at this point that the reader is wondering why we went through the trouble of assuming an approximate position and correcting our clock accordingly at the second hand level. We did not need this information to solve for user position and receiver clock error. The reason for the assumptions is twofold. First in the real GPS system it is really nice to know the approximate time so the receiver can start forming estimates of where the user is. By contacting just one SV the user receiver (at the earth's surface) can estimate GPS time to within 20 ms. This follows from the min/max known distance. Second by setting the receiver clock, of which there is a "second" counting Dial we have not yet discussed, the receiver can reduce the computations it takes to compute the true clock error. This follows from the fact that the 3-D solution of user position and clock error is an iterative process.

Physically correcting the receiver reference clock dials (or the electronic equivalent) is optional, as we have just discussed. But there is a use of GPS that needs this has to be done. Many GPS applications need GPS-supplied accurate *time* information. By using the calculated Tbias term to correct the dials below the 1 s dial level, we can provide a timing signal that is tied to GPS clock accuracy, which is extremely high. Since the receiver clock is usually a low-cost unit, the calculation of Tbias and receiver clock correction is performed frequently enough to keep the receiver's clock "honest." The exact rate of correction is dependent on the receiver's reference clock quality.

1.12   A Master Clock

GPS has a master clock. It is not a "physical" clock but rather a "paper" clock. It consists of calculations and measurements made by the Control Segment (CS) of the system. Up to this point the clocks in our linear model at points A and B, we have assumed to be in synchronism. We will now allow them to have a small error with respect to each other.

Figure 1.6 shows the new system. It is identical to the two-clock system we just covered except we have added a master clock. The camera still records the state of the two replica clocks and the receiver reference clock @ snapshot instant for use in

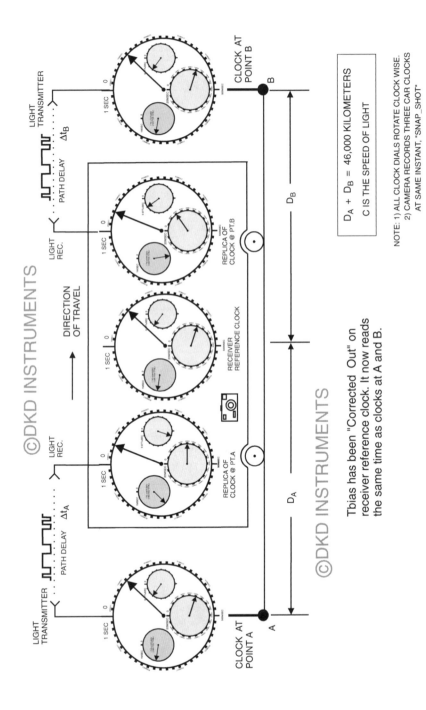

Fig. 1.5 Car clock has error (Tbias) corrected out using measured distance and times from clocks A and B

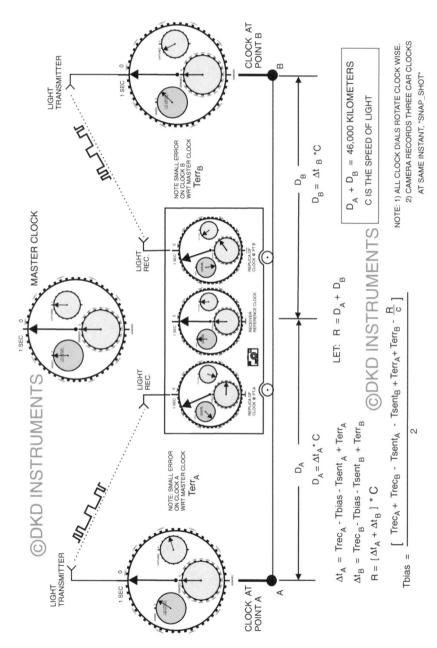

Fig. 1.6 Clocks at A and B have small error with respect to each other; each clock sends receiver its error term with respect to master clock

the calculations of Tbias and the distance to A or B. There is a difference: note that Clocks A and B are not synchronized below the 1 ms level. Inspection of the dials in Fig. 1.6 reveals this. The error is less than a millisecond. The receiver reference clock is shown synchronized with the master clock. How can the receiver reference clock be synchronized to our new master clock? The "trick" is that the clocks at points A and B "know" their respective errors as compared to the master clock. In the GPS the CS sends each SV its error with respect to the master clock. In our linear model the transmitters at A and B send their error terms (as part of the data stream) to the receiver. These terms tell the receiver how much the clocks at A and B they are "off" from "master" time. We can include these terms in our correction terms for the receiver's reference clock as shown in Fig. 1.6. The error terms must also be applied to receiver's replica clocks. In addition, if we *added* the path delay associated with each replica clock, the replica clocks would be in synchronism with the master clock. The receiver would still need to solve for its Tbias as before, but now displayed time it is referenced to the master clock time.

In this fashion the receiver's reference and replica clocks are adjusted so that the entire system is now referenced to the master clock. The reasons for using this method lie in the complexities associated with synchronizing multiple SV clocks. It is easier to allow a small error to exist and have the SV send the error information to the user receiver.

A final note on the master clock. As shown in Fig. 1.6 it has 20 tics on the 0.977 μs dial. In the GPS master clock this resolution is much higher. For all practical purposes this dial becomes an analog dial (infinite number of tics) and its precision is so high.

1.13   A "Second" Counting Dial for the Clock (Modified TOW)

The basic time unit of GPS is the second. Our clock model to date could measure up to 1 s of time with great accuracy. As we have said this is more than adequate for many receiver functions. But there are operations we will need to do later that need to know how many seconds have elapsed. Figure 1.7 shows our upgraded clock. It still has our four-Dial clock in its center but a "second-counting" Dial has been added on the outermost Dial. The second-counting Dial counts the number of seconds in 1 week, which is 604,784 s. It then restarts at zero and starts again. That makes the tic size of this Dial exactly 1 s. Every time the 1-s dial completes a rotation the second-counter dial increments by one tic.

The second-counter dial is also aligned with a day of the week. The rollover from 604,784 to zero happens at midnight of Saturday each week. Normally the second-counter is not implemented in hardware. The receiver computer would fulfill this function.

The GPS fundamental time unit is the second. As we see later, GPS has a counter called TOW (time of week) that does not explicitly count in seconds, instead it counts 1.5 s intervals. Therefore, there are 403,200 1.5-s intervals in 1 week.

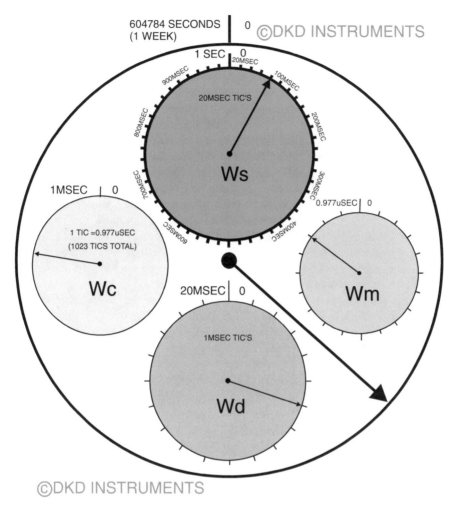

Fig. 1.7 Clock model with "seconds" counting dial added. Large dial counts 1 week worth of seconds which is 604,784

The rollover point is still Sunday at midnight. We will see that we can still properly set our second-counting Dial (in the receiver's replica clocks). This is possible as the GPS data stream contains timing marks that are on even multiples of 1.5 s, specifically there is a 6 s timing mark which is aligned on a (virtual) 1 s timing mark.

The author faced a choice of sticking to the TOW counter and in this process introducing a clock Dial that is asymmetric with the Dials below or adopting the more natural 1 s tic size for this Dial. There is good reason to use the 1-s counting Dial. Later we will see that in the calculations for SV position the time variables are scaled in seconds, not 1.5 s. This is an unfortunate complication but it must be lived with.

1.14   Time Tag the Pictures

The new second-counter dial was introduced here as we have a use for it. Let us take a look back at our model of Fig. 1.2 as now shown with a modification to the camera in Fig. 1.8. The camera has been triggered by a signal or event we have called snapshot. This time could occur at any time during the car's travel on the roadway. We can look at the photo and from it we can determine the sent and received times. The photo must show the replica clocks used and the receiver's reference clock for a valid measurement. There is another way to capture the reference clock information. In Fig. 1.8 we have tied the camera trigger button to the 1-s dial of the clock. The large second-counter Dial we just discussed is not shown for clarity, but its there. The effect of tying the camera to the 1-s dial is to take a photo every 1 s. The photo can now just show the replica clocks and need not show the receiver's reference clock as we have "Time Tagged" the photo. On the back of the photo we can record the reading of the large second counting dial. So the photo records the state of the replica clock(s) which is the time sent and on the back the "time tag" records the time received.

What is the purpose of "time-tagging" the photo when the method of recording the reference clock worked fine? It is difficult to "see" from this point in our discussion but this different approach to recording receive time has repercussions in the way the receiver implements its reference clock. The method of recording the reference clock with the replica clocks (All Clocks method) allows the capturing of time received and time sent at *any* instant we choose. The time tag method is more restrictive. The "All clocks" method requires a complete, fairly complex reference clock to be constructed. The time tag method allows the receiver's reference clock to be simplified considerably. If the aim of the receiver is recover GPS time than a more complex reference clock will be needed as well as the added mechanism to do the Tbias correction.

1.15   The Subtle Problem of Delays at the Receiver

In our roadway example a subtle assumption was made but not talked about. Let us return to that example for a moment. The assumption that we made, but did not bring up, was that the receiver in the car did not have any delay from the time the light pulse hits its receiving element to the moment the TOA is recorded. All practical receivers have delay in them. The question is what effect will this have on our TOA measurement and consequently our computed distance to point A (or B)? If the delay is less than 48 ns we would see no effect or degradation of our TOA measurement (i.e., it is below our clocks resolution). But let us assume that the delay is 1 ms. This is substantial and must be addressed, but how? We could try and measure the delay in the receiver paths. But there is a better way. If the delay is the same (or within a fraction of a microsecond) for light received from point A as from

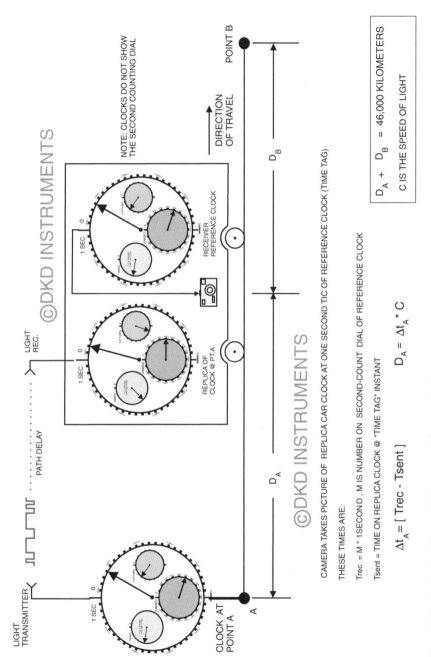

Fig. 1.8 Using the time Tag method to record time received

point B, then this delay will be *automatically* corrected when the car clock bias is solved for. In essence, any uniform delay in the receiver for light received at A or B becomes equivalent to a reference clock offset or bias.

1.16   Extending Position Measurement to 3-D Space

The roadway example is a one-dimensional model of the problem of finding distances and a receiver clock offset using multiple clocks. If the dimension were increased to 2-D, a plane, a third point would be needed to resolve distances and clock offset. At 3-D we need still another point to determine all unknowns. This is the situation with a user wishing to find position using GPS. A minimum of four satellites with known positions is required, one user position and user clock error. At first glance the line model (1-D) does not seem to have position information in it. But if one assumes point A at zero then point B is at the roadway length, which was given. So the presence of position information allows the solution of the user clock offset or bias and the distances to known points.

1.17   Summary

In this chapter we introduced the reader to the fundamental problem in GPS receivers: Synchronizing multiple clocks so that path delay can be measured. We constructed a highly accurate clock for making precision time measurements. The concept of transferring a clock time from one location to another was introduced (replica clock) as well as the effect of path delay on its displayed time. The replica was shown to contain the time of transmit of the signals we used. By taking a photo of reference clock and the replica clocks we could determine distances (path delays) and reference clock errors when we added a second clock to the linear model. The concept of a master clock was discussed as well as how to synchronize to it. Lastly we updated our clock with a second counting dial and discussed the concept and implementation of time-tagging.

Chapter 2
Introduction to the Global Positioning System

In Chap. 1 we introduced some of the working principles of the GPS by analogy with a simple linear model. Now its time to move into more detailed discussion of the GPS satellite system and how it works. In this chapter we will cover the basics of how the satellites are configured, making the path delay measurements, a quick look at the user position solution equations, and a high-level block diagram of GPS receiver will be presented.

2.1 The Satellite System

GPS is comprised of 24 satellites orbiting the earth at a distance of 20,000 km as measured from mean sea level. See Fig. 2.1. Figure 2.2 shows a more detailed picture of a user getting position information from four satellites. Each satellite contains a very precise clock. All the clocks and hence all the timing signals associated with each satellite are in near-perfect synchronism. The basic principle of operation is similar to the Roadway Distance Measurement example of Chap. 1. A rough model is to imagine a system of satellites emitting a "pulse" of light at exactly the same instant as all the others (Important: GPS is NOT a pulsed system). If you were at the exact center of the earth and the satellites were in perfect circular orbits (GPS orbits are nearly circular) these pulses would all arrive at your receiver at the exactly same instant. For all other points the pulses would (typically) arrive at different times. If one has knowledge of the positions of the satellites and can measure the path delay (distance) from at least four satellites to the user receiver, the user position X, Y, Z in ECEF (Earth Centered Earth Fixed) coordinates and user clock bias can be solved for.

D. Doberstein, *Fundamentals of GPS Receivers: A Hardware Approach*,
DOI 10.1007/978-1-4614-0409-5_2, © Springer Science+Business Media, LLC 2012

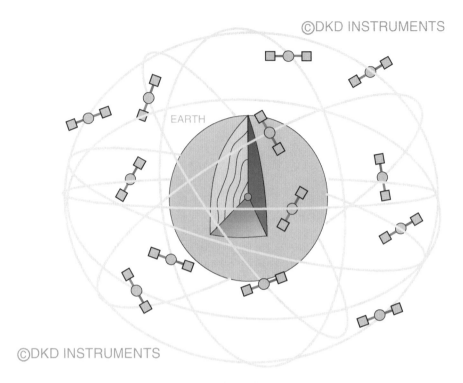

Fig. 2.1 The earth and its constellation of GPS satellites

2.2 Physical Constants of a GPS Satellite Orbit That Passes Directly Overhead

Figure 2.3 shows a simple model of one satellite circling earth directly overhead at the user zenith. We will use this simple model to compute some constants associated with a user at or near the earth's surface. We will assume that the user can be at a maximum altitude above sea level of 10 km. The minimum altitude will be assumed to be sea level. With this information and the known mean diameter of a GPS orbit we can calculate the minimum and maximum distance to the satellite as it passes overhead. Figure 2.3 shows the elements used in this calculation and other results. From this information we find that the maximum distance a GPS satellite is 25,593 km. The minimum distance is found to be 20,000 km. These distances correspond to path delays of 66 and 85 ms, respectively. As we have seen in Chap. 1, we can use this knowledge of maximum and minimum distance to set the receiver's reference clock to approximate GPS time.

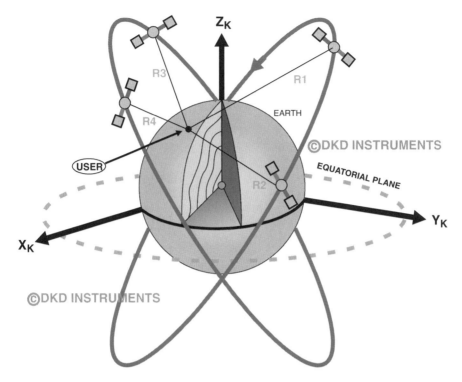

Fig. 2.2 User at earths surface using four SV's for position determination

2.3   A Model for the GPS SV Clock System

Each GPS SV has its own clock that "free runs" with respect to the other clocks in the system. GPS uses a "master clock" method in which *all* the clocks are "referenced" to the master clock by the use of error terms for each SV clock. We discussed this already in Chap. 1. In this chapter we will use the clock model of Fig. 1.7 *minus* the outside second-counter dial. In addition to the omitted second-counter dial, we will not show the dial that indicates the week of the year, year dial, etc. We have not discussed these new dials but they are present in the time-keeping method employed by GPS. For the purpose of discussion and understanding GPS we will often use the clock model of Fig. 1.7 minus the dials above the 0–1 s time increment. If we show or discuss a SV or Receiver replica clock without *all* the dials, the reader will assume the missing dials are "present" but not shown for reasons of clarity.

The smallest time increment dial of our clock model will always be the 0–0.977 μs dial. As mentioned in Chap. 1 this dial has a very fine "effective" resolution as used in the SV. But this statement is an approximation. In reality the 0–0.977 μs dial is the one dial of all the clock dials in our model that does not have a direct physical counter part in the "true" SV clock. In other words, our clock model at the sub microsecond level is not a completely accurate model of the GPS clock.

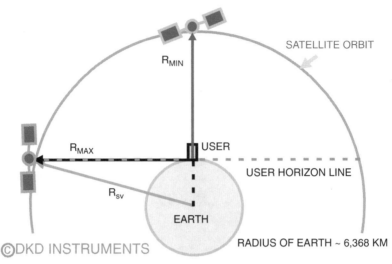

Fig. 2.3 The range from the receiver to a GPS satellite and physical constants (for a GPS satellite orbit that passes directly overhead)

When we use the clock model to describe the receiver's reference clock (or SV replica clocks) we will see that the 0–0.977 μs dial does have a direct physical counter part in the receiver. The impact of having the smallest time increment dial not an exact model for the SV clock is not an important issue for this text as our goal is position accuracy of ±100 m.

2.4 Calculating Tbias Using One SV, User Position Known

Perhaps the simplest application of GPS is synchronizing a receiver's clock when the receiver's position is known. In this example we will form an first-order estimate of the Tbias term associated with receiver clock. Figure 2.4 shows a model of our example system. The SV has its own clock and sends its clock timing information to the earth-based GPS receiver using an encoded radio wave. The receiver has two clocks, a reference clock and a replica of the SV clock reconstructed from the received radio wave. Figure 2.4 shows the receiver clock corrected with the calculated Tbias term, i.e., it is in synchronism with the SV clock.

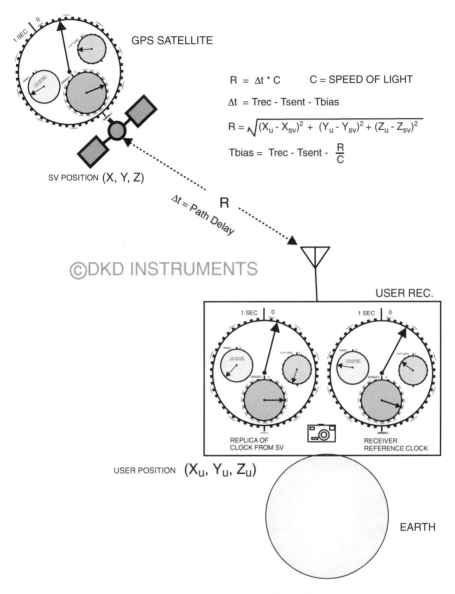

GPS SATELLITE

$R = \Delta t * C$ $C = $ SPEED OF LIGHT

$\Delta t = $ Trec - Tsent - Tbias

$R = \sqrt{(X_u - X_{sv})^2 + (Y_u - Y_{sv})^2 + (Z_u - Z_{sv})^2}$

Tbias = Trec - Tsent - $\dfrac{R}{C}$

SV POSITION (X, Y, Z)

$\Delta t = $ Path Delay   R

©DKD INSTRUMENTS

USER REC.

1 SEC   0   1 SEC   0

REPLICA OF
CLOCK FROM SV   RECEIVER
                       REFERENCE CLOCK

USER POSITION (X_u, Y_u, Z_u)

EARTH

Fig. 2.4 Calculation of Tbias using a single SV, user and SV position known

In order to solve for Tbias we need to calculate the distance R from the SV to the User receiver and measure Trec and Tsent as shown in Fig. 2.4. The calculation of R uses the distance equation which requires the user and SV positions in X, Y, Z. We have assumed the user position is known. In addition to the SV clock information the SV position information is encoded onto the radio wave that is transmitted to the receiver. We will assume for now that this information is provided in X, Y, Z coordinates. We will see later that getting SV position information in X, Y, Z format is nontrivial.

The Tsent and Trec information are obtained as before by just taking a "snap shot" of the receiver reference clock and the receiver's replica of the SV clock. If we record the SV position at the same instant that we record Trec and Tsent, we will have all the information needed to solve for Tbias. It is important to realize that due to SV motion we must "capture" the SV position data at the same moment we capture the state of the receiver's clocks. If we do not properly capture SV position, Tsent, and Trec, then the computed distance, R, would be incorrect for the measured path delay.

Now that we have all the information needed we can calculate R, Δt, and Tbias. If we continually update the measurements and calculations, the receiver reference clock will "track" the SV clock. This allows the GPS time receiver to replicate the stability of the SV clock. SV clocks are atomic based and so the stability is very high. If we modify the receiver's reference clock to output a "pulse" every time the 1 s dial passes the 0 tic mark and a 1 pps signal will be generated. This is a common signal many GPS receivers provide.

2.5 GPS Time Receiver Using Master Clock and the Delay term Tatm

In the previous example we computed Tbias when the user position was known. In this example we will include the effects of SV clock error with respect to the GPS master clock and the additional delay caused by diffraction of the radio beam as it passes through the earth's atmosphere.

Figure 2.5 shows the details of our new model. The receiver's reference clock is shown corrected to the master clock time. The SV clock has a small error with respect to master clock. The error is less than a millisecond. As before we capture the state of the receiver's reference clock and the replica clock. This information is used in conjunction with the computed path length R to form our estimate of Tbias. There is a difference from our first example and that is in the path delay. The expression for the path delay now has two additional terms. One, of course, is the SV clock error with respect to the master clock, Terr_sv. The other is the term Tatm.

The term Tatm is the extra delay experienced by the radio beam as it travels through the earth's atmosphere. To a first approximation the atmosphere acts like a lens and "bends" the radio beam from the SV to receiver. This bending of the radio beam as it passes through the atmosphere causes the extra delay. Normally the delay Tatm is broken into two parts, one for the Ionosphere and one for the Troposphere layers of the earth's atmosphere. Here we have combined them into one term with the sign convention following most of all GPS literature. How big is the added

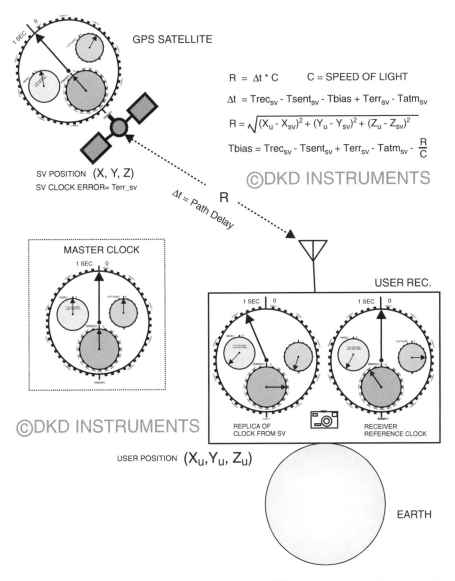

Fig. 2.5 Calculation of Tbias using a single SV, user and SV position known with master clock and atmospheric delay term

delay? Expressed as an increase to the R term this "extra" distance is in 50 m range. The maximum value occurs when the SV is low in sky and decreases as it rises. This delay is relatively large and consequently places the error introduced by Tatm at the top of a long list of error sources. Estimating the value of Tatm requires knowledge of the SV position in relation to the receiver, exact time of day, sun spot activity, day/night, etc. Needless to say, the effects of the Troposphere and Ionosphere on a radio beam are extremely complex and beyond the scope of this text. The reader is directed to the bibliography for suitable references.

By including the master clock correction and a term for atmospheric delay we form a model that achieves greater accuracy than our previous model. We also put into place the primary components needed to do the full solution to the user position problem. If we were to look at a 1 pps signal generated from the receiver's reference clock it "track" the GPS master clock with greater precision.

This example also illustrates that by refining the path delay estimate we can achieve greater accuracy in the computed value of Tbias. This will also be true for computed user position. It is hoped the reader sees the pattern emerging of successive refinements to the path delay, which results in greater accuracy for both position and time measurements.

2.6 Solving For User Position Using Four Satellites

With the previous examples and models we have laid the groundwork for the task of determining user position, Xu, Yu, Zu in ECEF coordinates and the user clock error, Tbias. Figure 2.6 shows the system using four SV's and four SV clock replicas in the user receiver. We again assume the user receiver is at or near the surface of the earth (i.e., each path delay is 66–85 ms). The four SV clocks are referenced to a master clock. The receiver's reference clock is shown with its error removed. In other words the receiver reference clock is displaying master clock time. For each SV to user distance a path delay is measured/computed. Additionally a distance R_i is assigned to the equivalent distance corresponding to path delay multiplied by the speed of light.

Figure 2.7 shows the set of equations. Each distance, R_i, can be computed from two different methods. The first method uses the path delay, which contains measured, computed, and SV-supplied terms. Each path delay has its own unique $Tatm_{sv\_i}$ term derived from estimates of user position. As user position estimates are refined, this term will also be refined in accuracy. The clock error for each SV is sent to the receiver and is identified by $Terr_{sv\_i}$.

Normally the Trec time would be the same for all the path delays. It is possible to have separate Trec times if the user receiver does not move appreciably from one Trec time to the next. As before we take a "snapshot" of the receiver's reference clock and the replica clock or clocks to record Trec and $Tsent_{sv\_i}$. If we choose to include all the clocks in our picture, then the value of Trec could be the same for all SV path delay equations. This last point needs a bit more clarification. If the user receiver is stationary (or moving slow compared to the time to measure each path delay) then we can measure the path delay to each SV separately. That is, we could first do SV_1, then SV_2, etc. In other words, a sequential measurement using a single channel receiver. As long as the drift rate of the receiver's reference clock is low enough over the time all four measurements are made, this method will work fine. Many early GPS receivers used this method as they were single channel receivers.

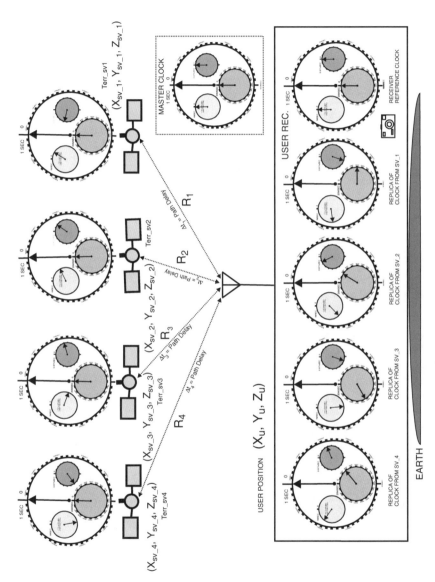

Fig. 2.6 Calculation of user position and Tbias using four SV's

Fig. 2.7 Equation set for solving for user position and Tbias using four SV's

$$R_1 = \Delta t_1{}^* C$$

$$R_2 = \Delta t_2{}^* C$$

C = SPEED OF LIGHT

$$R_3 = \Delta t_3{}^* C$$

$$R_4 = \Delta t_4{}^* C$$

$$\Delta t_1 = Trec - Tsent_{sv\_1} + Tbias - Terr_{sv\_1} + Tatm_{sv\_1}$$

$$\Delta t_2 = Trec - Tsent_{sv\_2} + Tbias - Terr_{sv\_2} + Tatm_{sv\_2}$$

$$\Delta t_3 = Trec - Tsent_{sv\_3} + Tbias - Terr_{sv\_3} + Tatm_{sv\_3}$$

$$\Delta t_4 = Trec - Tsent_{sv\_4} + Tbias - Terr_{sv\_4} + Tatm_{sv\_4}$$

$$R_1 = \sqrt{(X_u - X_{sv\_1}) + (Y_u - Y_{sv\_1}) + (Z_u - Z_{sv\_1})}$$

$$R_2 = \sqrt{(X_u - X_{sv\_2}) + (Y_u - Y_{sv\_2}) + (Z_u - Z_{sv\_2})}$$

$$R_3 = \sqrt{(X_u - X_{sv\_3}) + (Y_u - Y_{sv\_3}) + (Z_u - Z_{sv\_3})}$$

$$R_4 = \sqrt{(X_u - X_{sv\_4}) + (Y_u - Y_{sv\_4}) + (Z_u - Z_{sv\_4})}$$

The drawback is a better reference clock is needed as the method is relying on Tbias being approximately constant over the entire measurement sequence of the four SV's path delays. Being able to "measure" four Trec and four Tsent times *simultaneously* requires a four-channel receiver. This is the assumption made when writing the equations of Fig. 2.7.

The distance R_i can also be computed by using the distance formula as we did in the single SV example for obtaining Tbias. Each distance is computed from the user's position Xu, Yu, Zu and the position of the SV, which is assumed to be sent from SV as X_{sv_i}, Y_{sv_i}, Z_{sv_i} coordinates. By using the distance equations combined with the path delay equations we have enough information to determine the receiver's position and the receiver clock error. Unlike our simple solution for Tbias using one SV, the equation set of Fig. 2.7 *cannot* be solved in closed form for Tbias and Xu, Yu, Zu. The reason is that the equation for distance contains the square root function. This introduces a nonlinearity which precludes a "closed form" solution. To solve the equations a iterative method is employed. An iterative solution for user position and Tbias is presented in Chap. 5.

2.7   The Pseudo-Range

When we subtract the Tsent reading of SV from the Trec reading we are computing a *measured* path delay. But our measured path delay has errors. Tbias is rarely, if ever, exactly zero. Because of this error the distance obtained by computing the range from known user position and known SV position *plus* the distance equivalent of the user clock error, Tbias, is called the "Pseudo Range." This statement reflects the fact that this range is not exact because it has the error of receiver clock bias error expressed in it. In the terms used in this text and following conventional polarities of error terms, the Pseudo-Range for any given SV_i would be:

$$\text{Pseudo - Range}_i = [(x_u - x_{sv\_i})^2 + (y_u - y_{sv\_i})^2 + (z_u - z_{sv\_i})^2]^{1/2} + \text{Tbias} \times C \quad (2.1)$$

A related measured path delay version/estimation of the Pseudo-Range is:

$$C \times \Delta t_i \geq C \times [\text{Trec}_{sv\_i} - \text{Tsent}_{sv\_i} + \text{Terr}_{sv\_i} - \text{Tatm}_{sv\_i}] \quad (2.2)$$

2.8   A Simplified Model of the GPS Receiver

Figure 2.8 shows a high-level block diagram of a GPS receiver. The receiver shown has four channels. Only one of the four channels is fully detailed. The other three are shown as repeated blocks. Each channel has an identical SV clock replication block. This block replicates the SV clock from the received SV RF signal. Only one receiver reference clock is needed. The RF carrier is typically converted to a lower frequency and then sent to each channel.

Dividing down Receiver Reference Oscillator (RRO) forms the replica clock and the receiver reference clock. The receiver reference clock as shown can be identical in construction to the receiver's replica clocks or it could be of different construction. Later in Chap. 6 we will introduce an alternate reference clock based on the tagged SNAP_SHOT method. This approach can result in a simpler reference clock implementation. The reference clock reset function shown is used to start the clock at predetermined time or to "subtract out" Tbias from the reference clock so that it displays Master Clock time (within the errors of receiver). The receiver's computer can force Tbias to minimum by dynamically loading N_0, N_1, N_2 and N_3 values into the receiver reference clock registers. In some receivers this is optional and the error is held in software.

It is important to understand that the SV Replica Clock is never truly recovered. We can only form better and better estimates of the true SV clock as the receiver design is advanced. This same comment holds for the receiver's reference clock. We can never get to the point when Tbias is exactly zero, the goal is to drive it to a minimum.

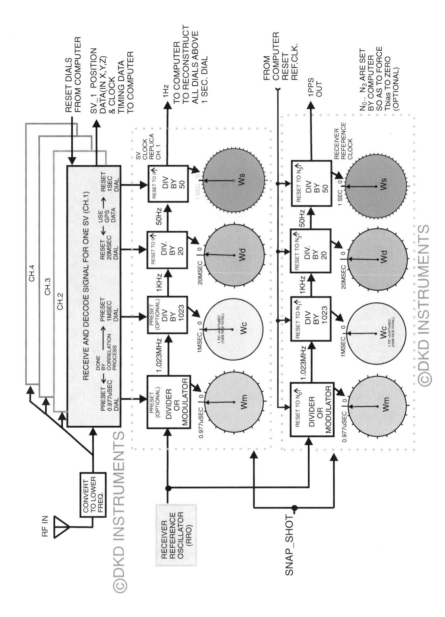

Fig. 2.8 Block diagram of four-channel GPS receiver

The SV replica clock must be properly reset in order to be a "faithful" delayed reproduction of the received SV clock. In the reception process the first two dials are correctly set by the *correlation* process used on the SV signal. In practice, the 1 ms dial is set by a quasi-dynamic process while the 0.977 µs dial is dynamically controlled to maintain "lock" on the received SV clock. The Preset inputs on these two dials is optional as some receivers do not "preload" values here. Since the output of the 0.977 dial drives all the other dials locking this dial to the received SV clock "locks" all the subsequent dials. The 0.977 dial may be a fixed divider in combination with a modulator or a pure analog method, which would use a separate reference oscillator from the receiver's reference clock (not shown).

The 20 ms and 1 s dials must be set by examination of the received data from the SV for that channel. Once the dials are initially set to the correct initial value they "free run" except for the 0.977 dial, which is under a servo loop control to keep lock on the received SV clock. Each channel must independently set its dials as per the method just described. After all channel clock dials are properly set then a SNAP_SHOT can be taken which records the state of all replica and reference dials (details of capture method not shown).

The data captured from replica and reference dials is now sent to the computer. In addition, each channel decodes the SV position information for the time the signal left the SV (i.e., Tsent). The computer now has SV position, Trec, and Tsent for all four channels. Now the computer can solve the equations of Fig. 2.7 (by iteration methods) for user position and receiver clock bias.

In the hardware section of the text we will get into the details of exactly how all this is accomplished. In Chap. 3 we will see how the transmitter, or SV, clock is constructed.

2.9   The Receiver Reference Oscillator

The reference oscillator used in the GPS SV is an atomic-based system or "atomic clock." Atomic clocks are extremely accurate and stable over long periods of time. Due to reasons of cost and size the RRO used in "lower end" GPS receivers cannot be an atomic clock. Typically, lower-cost GPS receivers use oscillators accurate to ±0.5 ppm, while the SV atomic clocks will exceed ±0.00001 ppm accuracy.

To investigate the effect of RRO error suppose we reset the receiver's reference clock such that it reads exactly 0 at the same instant as the GPS Master clock. We then let the reference clock "free run." After a period of time (an interval dependent on the PPM rating of the RRO) our reference clock would typically drift from the GPS Master clock resulting in a difference of displayed time on the dials of the two clocks. This occurs because our reference oscillator has a static frequency offset and drift rate. The result is the receiver reference clock will lose synchronism with respect to the Master clock as time advances. Once the error grows to approximately 1 µs we would see an equivalent distance error of 300 m in our path delay measurements. In general, higher-quality GPS receivers use higher-quality reference oscillators in order to mitigate the problems associated with RRO errors.

2.10   Satellite Position Information

In our model we assumed that the *XYZ* data was sent down to receiver on an encoded RF signal. The format was assumed to be the satellites position expressed in ECEF coordinates, *X*, *Y*, *Z*. Unfortunately it is not that simple. Rather than sending its position in *X*, *Y*, *Z* coordinates each SV sends a very complex set of *Orbital Parameters* for its particular orbit. This set of orbital data for GPS is called the *Ephemeris* data.

Orbital parameters can be used to describe the precise orbit characteristics of any orbiting body about the earth. The parameters contain the diameter of the orbit, its deviation from circular, its angle w.r.t. the polar axis, etc. If you know the orbital parameters of an orbiting body precisely, you can predict where it is and where its going, provided you know what *time* it is.

The SV orbits are fairly stable and repeatable. So once you have the orbital parameters for all the satellites at a particular day and time you can compute their positions at a later date (not too much later) with a fair degree of accuracy, using the same set of orbital data. In fact, as we will find out later each SV sends down the orbital parameters for not only itself but for all the other SV's of GPS but with reduced precision (Almanac Data). This allows the receiver that has a successful contact with just one SV to obtain position information for all the other satellites. Together with the initial estimate of present GPS time this information can assist the receiver in obtaining its first estimate of user position.

Once the receiver gets the orbital data it now has the task of converting this data to *X,Y,Z* in ECEF. In order to do this accurately it must know what "time" it is. We can get this information from the Tsent data present on the receiver SV replica clocks. Once we have the Tsent time we can use the orbital parameters to work out the *X,Y,Z* position of the SV at the Tsent instant of time.

2.11   Summary

Its time to take a step back and examine where we are in our quest to understand the GPS receiver and to get our first-order solution of the user's position. The first two chapters have laid the groundwork for understanding the basic principles of the using synchronized clocks to measure distance by measuring path delay. In this chapter we have discussed some of the details of the satellite system. It is hoped that the reader can start to see the real complexity involved and now has a grasp on the basic principles and working mechanisms behind GPS.

The first two chapters were written on a lay basis so as to insure that as many readers as possible can get the basics of GPS. From here on some readers may not have the background to proceed. Those readers may need to consult other texts to get the needed background knowledge to understand the discussions ahead. The GPS signal structure will be the next topic.

Chapter 3
GPS Signal Structure and Use

The purpose of this chapter is to explain the elements of the GPS radio signal that will allow the receiver to solve for user position and receiver clock error. The GPS radio signal has a complex signal structure. Fortunately we do not need to address all of it, only the parts we need. By investigating a simplified model of the SV transmitter, we can focus on the essential elements of the GPS signal that address the specific information we will need to solve for user position and receiver clock error. Also in this chapter, a method is presented that allows the receiver to properly "set" the replica clock Dials above the 1 ms Dial using the GPS data stream. Finally, the effects of Doppler on the GPS signal and its repercussions at the receiver are discussed at the end of this chapter.

3.1 A GPS SV Transmitter Model

Figure 3.1 shows a simplified model of a single GPS satellite transmitter structure. The actual structure of the satellite transmitter is considerably more complex but for our purposes this model is adequate. It is hoped that many of the elements of our GPS transmitter model are now familiar to the reader. In many ways, the receiver and transmitter clocks are nearly identical in basic structure.

The SV clock forms the heart of the system. The four Dials from 0.977 μs to 1 s are explicitly shown. The Second-Counter Dial (TOW), Week Counter Dial, etc., are not shown. Rather these Dials are contained in the 50-bit per second data stream sent by the SV to the receiver. The SV clock has an error with respect to the Master Clock as indicated on the SV clock Dials. This is the same error term, Terr_sv, we used in our earlier equations for the path delay. The amount of the offset is less than 1 ms.

An atomic-based oscillator (ABO) at 10.23 MHz is the primary reference for the SV clock and other generated frequencies. This signal is divided or multiplied to get all other signals in transmitter. This means that all signals generated by the SV are *coherent*. When multiplied by 154 the RF carrier at 1,575.42 MHz is generated. This frequency is called the L1 frequency. The L1 signal is transmitted Right Hand

D. Doberstein, *Fundamentals of GPS Receivers: A Hardware Approach*,
DOI 10.1007/978-1-4614-0409-5_3, © Springer Science+Business Media, LLC 2012

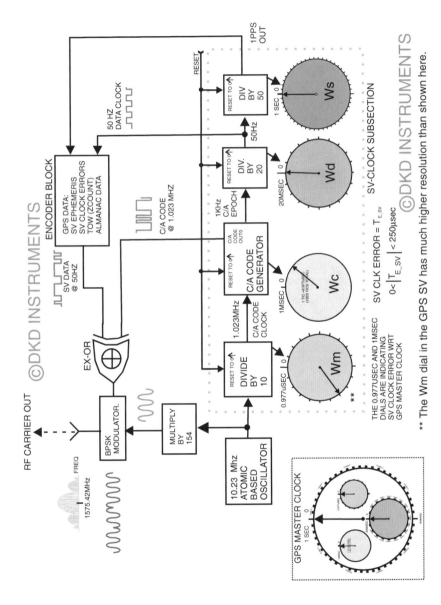

Fig. 3.1 Simplified model of a GPS satellite C/A code transmitter

circular polarized. There is another carrier called the L2 frequency not shown in Fig. 3.1. It can be used for two frequency measurements of the atmospheric delay term as discussed in Chap. 2. The L1 carrier is the most widely used signal so we will not address the L2 frequency further. Our focus in this text will be on receivers that receive only the L1 carrier. This is the most commonly available receiver on the market today.

The 10.23 MHz signal is divided by 10 in the SV clock subsection. The resulting signal at 1.023 MHz is used as the C/A code clock. This would translate into 10 tics on the 0.977 µs Dial. But the model shown omits some detail on the ABO. There would be additional finer controls for the frequency of this signal up to the 1.023 MHz point. As far as the receiver is concerned, the exact details are irrelevant. The reason is that the error term Terr_sv will correct the displayed time of the receiver-generated replica of the SV clock to be that of the master clock.

The next block is the C/A code generator. This block effectively does two operations at once. It produces a PRN sequence 1,023 bits long and in the process divides by 1,023. Examining our model of the receiver SV replica clock subsection of Fig. 2.8, this was shown as a fixed divider. This was a simplification and the receiver will use its own replica of the C/A code generator at the same spot as we see in our transmitter model. For readers not familiar with PRN code generators please see Appendix B. The C/A code repeats every 1,023 bits and when the repeat time is decoded the 1 kHz C/A Epoch signal is produced. The timing relationships between the C/A code, C/A Epoch, and 50 Hz data are shown in Fig. 3.2. The C/A code generator corresponds to the 0–1 ms clock Dial.

The 1 kHz C/A epoch signal is sent to the divide by 20 to produce the 50 Hz data clock. This corresponds to the 0–20 ms Dial of the SV clock. The data clock is fed to Encoder block where all the SV information is assembled to send to the receiver. This is comprised of the SV ephemeris data, SV clock error, Almanac data, etc. The divide by 50 produces the 1PPS signal and the 0–1-s Dial. Now we can see that the 0–1-s Dial is counting data bit clocks. The 1PPS signal is fed to the encoder block where it is combined into the data to embed all clock Dials above the 0–1 s Dial. This would be the seconds in one week Dial (TOW), week of the year Dial, etc.

3.1.1 Embedded Timing in the 50 Hz Data

After the 50 Hz data are encoded, it is Exclusive OR'ed with the C/A code and bi-phase modulated onto the L1 carrier. The carrier is amplified and transmitted down to the receiver on the earth's surface. The crucial observation at this point is that the RF carrier *does not* contain the SV clock timing signals explicitly.

In particular, the 1.023 MHz, 1 kHz, 50 Hz and 1 PPS timing signals are not directly encoded onto the carrier. The only explicit information on the carrier is the C/A code and the 50 Hz data (not the data clock). This has direct repercussions at the receiver. The receiver must use the C/A code and the 50 Hz data to construct a replica of the received SV clock. In other words, the receiver must *derive* the

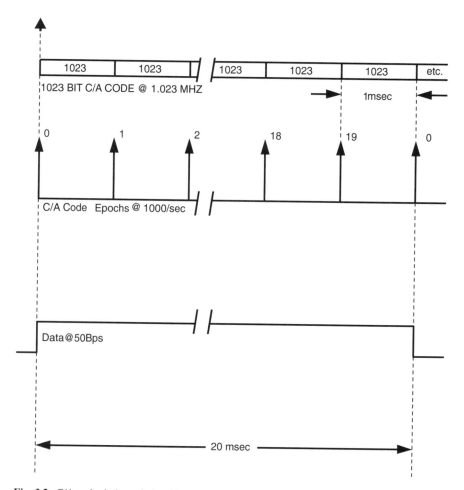

Fig. 3.2 C/A code timing relationships

missing timing signals from the received C/A code and the 50 Hz data that will be used to properly set all the Dials of the receiver replica clock(s) for any given SV. We will see that the correlation process will automatically set the 0.977 μs and 1 ms Dials. But the receiver replica clock Dials above this level must be set be using the timing "embedded" in the 50 Hz data. At the end of this chapter, we address some of the details of how this is accomplished.

3.1.2 BPSK Modulated Carrier

The L1 GPS signal is an RF carrier at 1,575.42 MHz modulated using the BPSK modulation method. BPSK stands for Binary Phase Shift Keying. Some details of BPSK can be found in Appendix C. All of the data information that the satellite sends

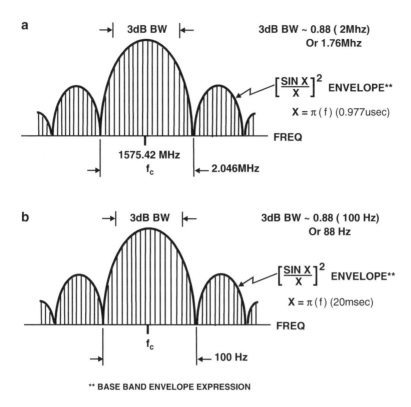

Fig. 3.3 (a) L1 power spectrum. (b) Data spectrum on carrier

to the user receiver is encoded onto the RF carrier using BPSK. BPSK modulation is a constant amplitude modulation scheme. This helps in signal power but creates other complexities, as we shall see. Figure 3.3a shows the power spectrum of the GPS C/A signal. The power spectrum envelope is approximated by:

$$\text{Power spectrum envelope of L1 signal} \approx \frac{[\sin^2 X]}{X^2} \tag{3.1}$$

where $X = \pi f(T)$ and $T = 0.977$ μs.

This formula would be exact if the C/A code were infinite in length. But since the C/A code repeats at 1 kHz intervals, a careful inspection of the C/A code spectrum would reveal a series of spectral lines separated by 1 kHz in frequency. The amplitude of each line will trace out (approximately) the $[\sin x/x]^2$ envelope. More detailed analytical expressions can be written for the power spectrum of the GPS signal. Regardless of their complexity, the reader is cautioned that complexity does not translate into reality. Issues of carrier leakage at the modulator and correlators, phase changes through media or signal processing elements, etc., all conspire to make an exact mathematical statement of the actual signal very difficult.

3.1.3   The Reset Line

Figure 3.1 shows a line labeled "reset." The reset function shown here is a simplified version of that in the real transmitter. The purpose of the reset line is to start all the dividers/counters that comprise the SV clock at roughly the same instant in time. This signal, when applied simultaneously to all SVs, forces all SVs to be transmitting the same C/A code bit, data clock edge, Data Preamble at *almost* the same time. Of course they are not transmitting these timing signal at *exactly* the same time due to SV reference clock errors and other imperfections in the reset process. So the reset line is used to synchronize the SV timing edges.

3.2   Virtual Time Alignment

In Fig. 3.1, we see the reference oscillator is an atomic clock at 10.23 MHz as its primary reference. This is a very accurate clock but even still it will drift with respect to GPS time as we have touched on earlier. This drift or error of each SV is kept track of by the ground control segment. As we now know these errors are computed and sent up to the SV. The clock error for each SV is unique to that satellite. The SV clock error data are sent down to the user receiver via the 50 Hz data stream. These data will allow the user receiver to determine these errors and correct them out of its reference clock. This method will guarantee that *all* SV time signals are aligned in time to a very high degree of accuracy. To explain that a bit more lets take another look at Fig. 3.1.

As shown in Fig. 3.1, all signals are derived from the 10.23 MHz atomic reference. This means that all signals inside the SV are *coherent* in time. This time alignment of the internal signals in each SV transmitter goes beyond an individual satellite: *The entire system of GPS satellites is in virtual time alignment*. Virtual time alignment is achieved by using the clock error data transmitted by each satellite as we just discussed. By correcting the errors of each SV clock at the user receiver, the entire system of satellites appears *virtually* coherent in time. This virtual synchronizing method is key to GPS and warrants a formal definition.

Virtual time Alignment Definition: *Once SV clock errors are properly subtracted out at the user receiver all GPS timing signals such as data clock rising/falling edges, code clock rising/falling edges, code generator reset state (C/A Epoch), etc., can be assumed to be sent (or occur) by all SVs at very nearly the same time or instant. There will be residual timing errors but their magnitude will be so small that they are not significant for the distance accuracies sought by this text and by many users of GPS.*

3.3   The C/A Code in GPS Receivers

The C/A code has a number of different uses in GPS. Let's list some of them:

- The C/A code allows us to identify from which satellite we are receiving data. Each satellite has its own unique C/A code. Essentially, the C/A code gives each SV an address. If we use this address properly we can determine from which SV we are receiving data.
- The C/A code allows the receiver to "lock" onto the satellite signal. The receiver does this by having its own copy of the C/A code used for each SV.
- The C/A code is used in recovering the replica clock at the receiver. By "locking on" to the transmitted C/A code, the receiver "recovers" some of the timing information associated with the transmitted clock information from a particular SV. Once the receiver "tracks" the C/A code, the *rate* of *all* receiver replica clock Dials is recovered. The C/A code track process also recovers the *phase* of the 0.977 and 1 ms Dials of the receiver replica clock. The *phase* of the receiver replica clock Dials above 1 ms is typically determined by using the GPS data stream.
- The C/A code "spreads" the RF carrier over a band of frequencies. At the receiver, this same spreading action helps defeat unwanted signal that attempt to enter user receiver. This property of the C/A coded carrier is sometimes called "Anti-jamming."
- The C/A code allows all SVs to use the same frequency to transmit on. For the receivers covered in this text this is the L1 frequency at 1,575.42 MHz. This fundamental property is called *Code Division Multiple Access* or CDMA for short.
- The C/A code-modulated carrier allows a pulse to be sent without reducing the carrier power. If we sent a pulse by turning the carrier on/off (pulsed carrier), we would reduce the average power of the carrier. Pulsed or ON/OFF carriers must have a much higher peak output power to obtain the same average power as the CDMA carrier. The use of CDMA techniques makes the satellite transmitter more efficient in terms of its size, weight, power consumption, life span, and cost.

3.4   Hidden Signals

Figure 3.4 shows the time waveform of GPS L1 signal carrier and its relationship to C/A code and 50 Hz data bit. We see that the carrier is never pulsed in the traditional sense. We only have data and the C/A code which are combined by EXOR and modulated onto the carrier.

Another important point is that data clock is derived from 20 C/A code repeats. Every data clock rising edge is synchronous with the rising edge of the first bit of the 1,023 long C/A code. All data edges are lined up with C/A code edges. The 1,023 bit C/A code will not have a $180°$ phase "flip" in the middle of any particular

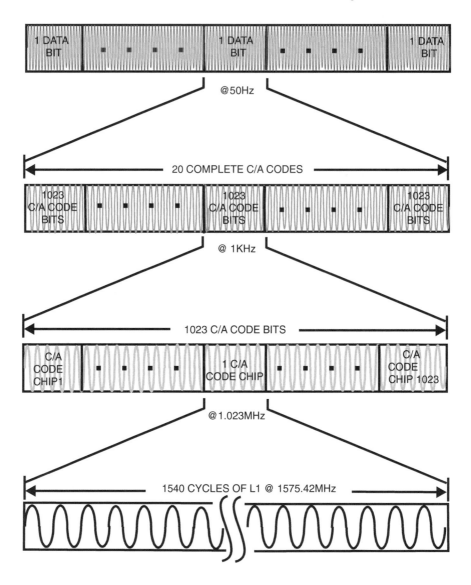

NOTE: CARRIER CYCLES NOT TO SCALE.

Fig. 3.4 L1 signal structure, from carrier to data bit modulation

C/A code cycle. The net result of the EXOR operation of data with the C/A code is to flip the L1 carrier phase another 180° per 20 C/A code cycles (at most), which will *not* show up on the time waveform. Essentially, the 50 Hz data are hidden in the L1 carrier in *time and frequency* by the C/A code.

3.4.1   *Data Hiding and Data-Modulated Carrier Spectrum*

A key feature of CDMA and GPS is to embed data into PRN stream, the C/A code in the GPS. As just discussed this is done with an EXOR logic operation. This embedding of the data into the C/A code completely hides the data from observers who do not know the C/A code. The only way to retrieve the data at the user receiver is to have a copy of the C/A code. Until this is done, the receiver will only see a random modulated carrier (in time) and power spectrum devoid of the data modulation portion. With a copy of the C/A code, we can "strip away" the code modulation on the carrier. This will result in a carrier with 50 Hz data modulation only. The signal now has a bandwidth of approximately 100 Hz. The power spectrum of the carrier with 50 Hz data is shown in Fig. 3.3b. The shape of the spectrum with just data on it is again approximated by the $\text{Sin}^2 x/x^2$ formula [$x = \pi f(20 \text{ ms})$]. This follows from the fact that the data are *nearly* random at 50 Hz BPSK modulation. The peak power of the data-only spectrum is higher by approximately 20 dB than the C/A-coded spectrum. This is because the power that was spread out over several megahertz has now been condensed into approximately 100 Hz by the C/A code removal process. The process of code "removal" (Correlation) is discussed in detail in Appendix A. Correlation increases the SNR dramatically inside the user receiver.

3.5   Received Signal Power by User at Earth's Surface

Figure 3.5 shows a plot of "signal power verses elevation angle" to the satellite as seen by the user at earth's surface. The signal power of the GPS signal is very low. The power indicated in the plot is the power of the carrier with no modulation on it.

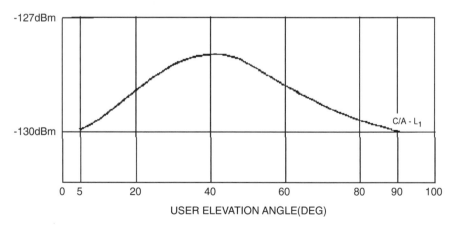

Fig. 3.5 L1 carrier power (unmodulated) in A1 HZ BW versus elevation angle for user at or near earth surface (3 dBi Lin. polarized ANT)

Once modulated the peak power is reduced as the signal power is now spread across approximately a 2 MHz bandwidth by the C/A code modulation. The C/A code allows all satellites to reuse the same frequency for receiving while spreading the carrier power over a band of frequencies. This spreading is good for protection against unwanted signals getting into the receiver but complicates trying to receive the C/A signal. In fact the signal power of the spread C/A signal is typically *below* the noise floor of all receivers. In effect the signal has a *negative* SNR. The ability to receive negative SNR signals is made possible by the CDMA method used to construct the C/A-coded carrier.

As we now see the C/A code plays a crucial role in rebuilding the SNR, obtaining code clock recovery and reconstructing the receiver replica clock(s). Without the C/A code, we cannot retrieve the GPS signal from the noise present in the receiver.

3.6   P-Code Receivers

The P-Code is a very long PRN sequence with a clock rate of 10.23 MHz. It is modulated onto the L1 carrier in 90° quadrature with respect to the C/A code. The P-Code is not shown in Fig. 3.1 The P-Code is primarily used by Military-type receivers. An interesting and important property of the P-Code is that even though it is present on the L1 carrier commercial C/A code receivers can ignore it. C/A code-only receivers are the dominant receivers sold today.

The P-code signal can be ignored for a C/A code receiver. The C/A code receiver does not need to know the P-code in order to obtain position information or the GPS data stream. This text does not specifically address P-code receiver design or function

3.7   GPS Data Structure Overview

In this subsection, we will introduce some the details of the GPS data stream or structure. What we intend to focus on here is the structure of the data and the retrieval of timing information needed by the receiver to properly reconstruct its copy of a particular SV clock. Some of details of the data stream are covered in Chap. 4 and also in Appendix D. The ICD-200 document should be consulted for additional information on the very complex GPS data stream.

A GPS SV sends down to the earth-based receivers a 50-bit per second data stream modulated onto the L1 carrier using BPSK. In the data block of Fig. 3.1, we see that these data are comprised of SV Ephemeris data (SV position), TOW (GPS time of week information), GPS Clock error information, etc.

The basic message unit is one *frame* of data which is 1,500 bits long (see Fig. 3.6). Each frame is comprised of five *subframes*. Each subframe is 300 bits

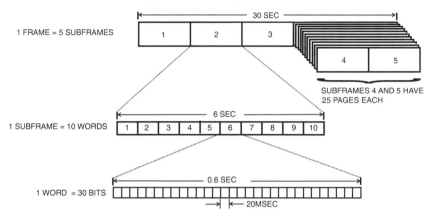

Fig. 3.6 GPS message format showing frame, subframe, word and master frame relationships

long. The subframes are then broken down into words. There are ten words per subframe, therefore a word contains 30 bits. Figure 3.6 shows the break down and the time of each element of the data message.

In addition, subframes 4 and 5 have 25 pages each. The entire set of five subframes and the 25 pages of the subframes 4 and 5 constitute a *Master Frame*. A complete master frame is sent every 12.5 min and is 37,500 bits long.

Each subframe starts with a TLM (Telemetry) 30-bit word then a HOW (Hand Over Word) 30-Bit word (see Fig. 3.7). The TLM word starts with a special sequence of 8 bits called the *Preamble*. The preamble allows the receiver to locate the start of each 300-bit subframe. Following the Preamble bits, the rest of the TLM word is comprised of the TLM message and parity bits. We will not pursue the details of the TLM message at this text. The information that we need to focus on here is the Preamble sequence. This sequence repeats every 6 s and the rising edge of the first bit of the preamble sequence is aligned with a (virtual) GPS 1-s timing edge.

The second word of every subframe is the HOW word. This word contains two important pieces of information. First is the TOW count message. The second is the subframe ID number that identifies which subframe we are examining. TOW is a counter that starts at zero on Sunday midnight and counts the number 1.5 s intervals in 1 week. Being that the last two digits are not included in the TLM message, the value given in first 17 bits of the HOW word has an LSB of 6 s. The point in time that this count is referenced to occurs on the first bit of the *following* subframe. Therefore, if the receiver has decoded this subframe it knows the exact value that the TOW count will be at the next rising edge of bit 1 of the next preamble sequence.

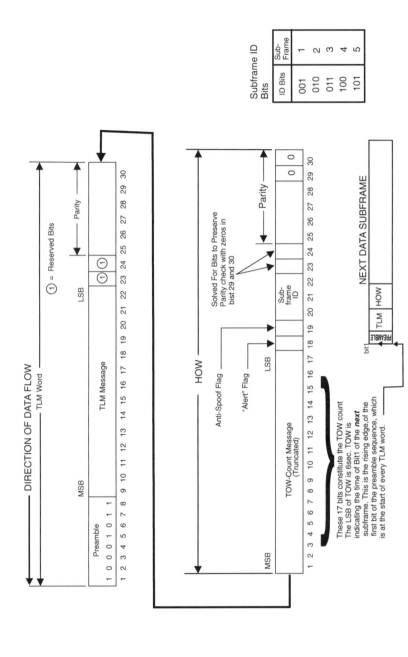

Fig. 3.7 TLM and HOW formats

3.7.1 Using the Data to "Set" Replica Clock Dials in the Receiver

We now have enough information to correctly set the remaining clock Dials of the receiver SV replica clock(s). We now know that the correlation process determines the rate of the entire set of receiver replica clock Dials but not the *phase* of the Dials above 1 ms. Figure 3.8 shows how we can use the GPS 50 Hz data stream to correctly set Dials above the 1 ms Dial and up to the Second-Counter Dial.

The 20 ms replica Dial can be properly set by resetting this Dial to zero at *any* rising data edge. This is true because all data edges are aligned in time with the 50 Hz data clock (20 ms period). The 1-s Dial can be properly set by resetting this Dial to zero at *any* rising edge of bit 1 of the preamble sequence. The preamble sequence always starts on a 6-s timing edge or boundary as it is the first bit of every 300 bit subframe (6 s period). The 6-s timing mark is an even multiple of 1.5 s (×4) so this puts the 6-s timing mark exactly on a (virtual) 1-s timing mark.

The second-counter Dial can be properly set by reading the TOW message and computing the number of seconds from midnight Sunday it represents. This is then applied to the second-counter Dial at the point in time of the next subframe bit 1. This timing mark is again aligned with a virtual 1-s timing mark as it is referenced to the 6-s subframe repeat rate.

By using the timing marks and the TOW data embedded in the GPS data, the clock Dials of the receiver replica clocks can be set to read the time they would display at the moment those signals left the SV. In other words, the receiver replica clocks will display the time sent information.

3.8 The Doppler Problem

For the user at or near the earth's surface, a GPS satellite travels with sufficient speed to induce appreciable Doppler on the RF radio wave. The Doppler can be predicted by using a vector approach and equations as shown in Fig. 3.9. In this figure the radial velocity component of the SV is broken into two parts, one perpendicular to the Line of Sight and the other parallel to the Line of Sight (LOS) vector. The component parallel to the LOS, V_d is the velocity component that causes the Doppler effect. As the SV passes through its highest point in the sky $|V_d|$ will pass through zero, which results in zero Doppler shift. It will also undergo a sign reversal, that is from a positive to a negative Doppler. Maximum Doppler is encountered when the SV is low in the sky, i.e., low elevation angles. The SV radial velocity is not constant in magnitude or direction. This is due to the very slight elliptical orbit and small forces that perturb the orbit.

The range of Doppler induced on the L1 carrier is approximately ±5 kHz. This frequency offset appears not only on the received carrier at 1,575.42 MHz but also appears on all the timing signals derived from the carrier. This follows from the

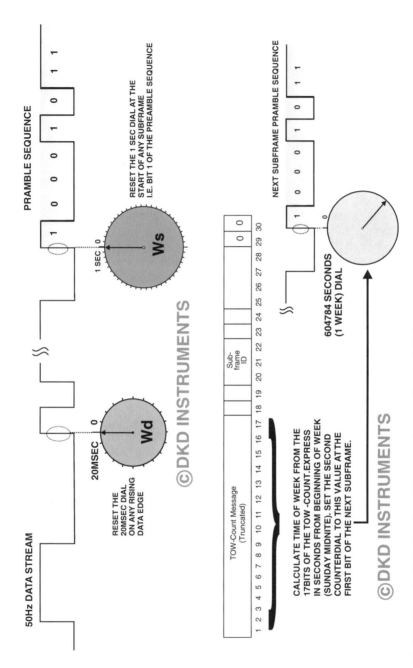

Fig. 3.8 Using the 50 Hz data from SV to properly set replica clock Dials

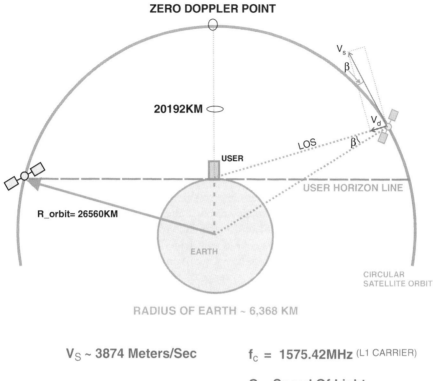

Fig. 3.9 Doppler calculations for a GPS SV, L1 carrier

coherent nature of the GPS signal structure, which derives all internal timing and frequencies from a common source. The exact amount of shift at a particular GPS signal will depend on the multiplying ratio between the carrier frequency and the signal in question. For example, the C/A code clock is at a nominal frequency of 1.023 MHz. The ratio to the carrier is exactly 1,540. Therefore, the C/A code clock as seen by the receiver could have a Doppler-induced shift of approximately ±3.2 Hz.

Doppler has a large effect on the design and implementation of the GPS receiver. In particular, the Doppler on the carrier translates to an uncertainty as to where exactly in frequency the carrier is during the signal acquisition operation by the receiver. The bandwidth of the GPS signal after the C/A code is removed (by correlation) is approximately 100 Hz. Conceivably, the receiver could use a 100 Hz filter in the signal processing of the GPS signal. The problem is that the induced Doppler will force the receiver to examine the entire 10 kHz span about the nominal carrier frequency to find the 100 Hz "slice" that has the GPS signal in it. The search range is usually larger due to errors in the receiver's own locally generated carrier used to convert the L1 frequency to a lower frequency. A 1,500 MHz carrier using a ±1 PPM reference adds roughly 3 kHz to frequency search span. It is tempting to make the receiver bandwidth wide enough to encompass the entire range of Doppler and local frequency offsets. But doing so would lower SNR in the receiver to unacceptable levels. The Doppler on the carrier forces the GPS receiver to implement some sort of frequency search method.

The Doppler also impacts the C/A code correlation process. The code clock Doppler offset of ±3.2 Hz seems small but it ensures the use of a servo loop to dynamically track the code clock at the receiver. The changing Doppler as the SV moves with respect to the receiver also requires the use of control loop to keep the signal inside the receiver's bandwidth. This joint requirement to search for the GPS signal in frequency and C/A code phase adds considerable complexity to the GPS receiver design. It is possible to produce a receiver that recovers the code clock by recovering/tracking the Doppler. This approach is not how modern approaches typically tackle these two problems.

3.9   Summary

In this chapter, we introduced the reader to basics of the GPS signal structure. A simplified model of the SV transmitter was presented. We used this model to show how the SV clock information is sent to the receiver as well as SV position. The basics of the GPS data message were introduced. We then discussed how the information in the data could be used to properly reconstruct the replica clock at the receiver. Finally, we discussed the effects of Doppler both on the transmitted signal and the repercussions of this Doppler on receiver operation and design.

Chapter 4
Solving for SV Position

In this chapter, we will solve for the SV position at the "time sent" moment. In order to do this, we will need to understand how to properly express the time sent information so that it can be used in the SV position equations. Also in this chapter, we discuss some details of the SV clock error correction.

4.1 SV Position

As we know from our earlier work, GPS does not send the SV position in convenient format such as X, Y, Z. Instead, it sends Ephemeris data that describes each SV orbit about the earth. The receiver uses the time sent information captured from the receiver's replica clock for a particular SV to determine where in its orbital path the SV was at the "time sent" moment. Once the receiver has the orbital parameters and a precise moment in time for that orbit, the position of the SV in ECEF (XYZ) coordinates can be calculated.

   The Ephemeris data are sent down in the 50 Hz data stream. The details of where in the data stream this information resides can be found in Appendix D. Our goal is to use the supplied orbital data to solve for the SV position in ECEF coordinates for any given "time sent."

4.2 Coordinate System

We have already been referring to the coordinate system that we wish to express the SV position in the Earth Centered Earth Fixed (ECEF) coordinate system. This coordinate system is "fixed" to the earth and rotates with it. It is defined by a standard called the World Geoditic System 1984 or WGS-84 for short. In this system, the Z_k axis of Fig. 4.1 passes through the earth's average pole of rotation.

D. Doberstein, *Fundamentals of GPS Receivers: A Hardware Approach*,
DOI 10.1007/978-1-4614-0409-5_4, © Springer Science+Business Media, LLC 2012

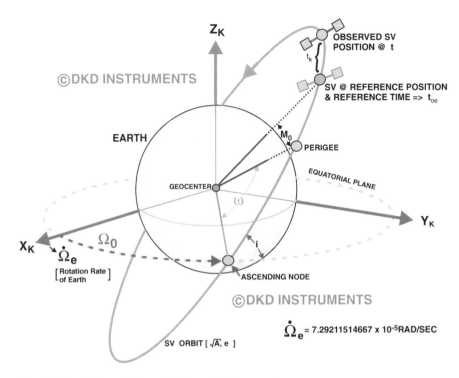

Fig. 4.1 Description of SV position from orbital parameters

The X_k axis passes through the intersection of the Greenwich meridian and the equatorial plane. The Y_k axis is $90°$ from these two defined axes.

4.3   Multiple Clocks, One Master Clock, and One Time Unit

A confusing element of solving for the SV position has to do with sorting out exactly what time scale we are referencing. When we speak of time in context of GPS we can chose SV time, GPS time, and User clock time. This statement reflects the fact that all these clocks are running at slightly different rates. But this leads to lots of confusion. There is really only one clock we care about, and that is the GPS Master Clock. We have used a number of names for this time scale: Master Clock, GPS Master Clock, or GPS Time. They are just different names for the same clock. The Master Clock uses the second as its basic unit and scale. Our goal is to make all other clocks (Corrected Receiver and SV Clocks) read this time as close as possible. We need to define a variable, that when we write it down, we know for sure that it refers to the time indicated on the GPS Master Clock:

When the variable "t" is used, it is in GPS time with units of seconds.

This definition of t also follows that used by the ICD-200°C document.

4.3.1 The SV Clock Correction Terms

Although slightly out of place here in our discussion of the SV position calculation, we need to digress and discuss the clock error terms which are contained in data subframe 1. As discussed many times, the SV sends down a "correction" term to receiver that enable the receiver to "correct" its replica clocks to the master clock time scale. The correction "term" is actually three parameters that model the SV clock's static "offset," drift and rate of drift with respect to the master clock. For our purposes we will only use the static offset term, which is called a_{f0}. This is the term we will use to correct the time indicated on the receivers replica clock(s). In other words, when we capture our replica clocks at the "snap-shot" instant they will have this offset with respect to the Master Clock. To reduce the SV clock error, apply the static correction term to the time sent:

$$\text{Corrected ``Time Sent''} = [\text{Replica Clock State @ ``Snap - Shot''}] - \text{Terr\_sv} \quad (4.1)$$

Where Terr_sv $= a_{f0}$

The impact this correction makes on computing the SV position is small. The reason is a_{f0} is small, less than a millisecond. Even at the speed the SV is moving the error induced in computed SV position by a 1-ms time error is not large. In 1 ms the SV moves about 4 m. But this is the radial distance the SV moves. The change in distance from the SV to the user would be far less. In short for our goals it is optional to "correct" the time sent data from the receiver's replica clock(s) to compute the SV position.

The effect on the computed path delay (Trec–Tsent) of the SV can be far larger due to the vastly larger speed involved, lights–speed. A 1 ms error in path delay translates to an error in distance of 300 km. For this reason, when the path delay is computed it is quite necessary to include the SV clock error term.

4.3.2 The Ephemeris Time Reference Variables t_{oe} and t_k

The orbits described by the ephemeris data are referenced to time variables t_{oe} and t_k. It is these two times that are inserted into the equations used to solve for the SV position. In other words, one cannot use the "time sent" information directly. Instead, computations and the current ephemeris data are used to arrive at the values t_{oe} and t_k we will need for the SV position calculations.

M_0	Mean Anomaly at Reference Time
Δn	Mean Motion Difference From Computed Value
e	Eccentricity
$(A)^{1/2}$	Square Root of the Semi-Major Axis
Ω_0	Longitude of Ascending Node of Orbit Plane at Weekly Epoch
i_0	Inclination Angle at Reference Time
ω	Argument of Perigee
$\dot{\Omega}$	Rate of Right Ascension
IDOT	Rate of Inclination Angle
C_{uc}	Amplitude of the Cosine Harmonic Correction Term to the Argument of Latitude
C_{us}	Amplitude of the Sine Harmonic Correction Term to the Argument of Latitude
C_{rc}	Amplitude of the Cosine Harmonic Correction Term to the Orbit Radius
C_{rs}	Amplitude of the Sine Harmonic Correction Term to the Orbit Radius
C_{ic}	Amplitude of the Cosine Harmonic Correction Term to the Angle of Inclination
C_{is}	Amplitude of the Sine Harmonic Correction Term to the Angle of Inclination
t_{oe}	Reference Time Ephemeris
IODE	Issue of Data (Ephemeris)
SIMPLIFIED SATELLITE VEHICLE CLOCK CORRECTION TERMS	
a_{f0}	**SV CLOCK OFFSET FROM GPS TIME**

Fig. 4.2 Ephemeris data definitions and clock correction terms

4.3.3   Ephemeris Reference Time, t_{oe}

This is one of the 16-ephemeris parameters sent (see Fig. 4.2). It is shown in Fig. 4.1 as the position of the shaded SV. The time t_{oe} is in the Master Clock time scale. The ephemeris data are based on a single point of reference in space and time. This reference occurs at a particular point in the defined SV orbit and it occurs at a particular point in time. That point in time is t_{oe}. When you calculate the SV position at other points on the orbit (other points in time), the method used is to extrapolate other SV positions from this reference position–time pair. Therefore as the SV moves away from this reference point in space (or equivalently time), the accuracy of prediction of the SV position will degrade. This reference time, t_{oe}, can be in the future or the past from current GPS time. Typically, it is updated hourly (By CS) so that errors of prediction do not grow out of tolerance.

Another way to explain what is occurring with the orbit parameters is to realize that, even though the parameters sent describe a *complete* orbit, typically only a small section of *that* particular orbit is used. This small section, whose center is located in time by t_{oe}, is further refined by correction terms that essentially modify the predicted path from the original primary (defined) orbital path. After approximately an hour, a whole new orbit definition is sent up from the ground control segment of GPS. The user receiver then repeats the SV position estimation process with new ephemeris and value of t_{oe}. The user receiver should always be using the "Freshest" orbital data set available for maximum SV position accuracy.

4.3.4 The Delta Time t_k

Figure 4.1 shows a term between the SV positions at GPS time t_{oe} and a SV position at a later time t. This is the term t_k. This term as shown in Fig. 4.1 could be wrongly interpreted as a distance. It is not. Rather t_k is a delta time between those two satellite positions shown. Specifically, t_k is a direct measurement of the amount of time between the current SV position at time t and the ephemeris reference time at t_{oe}. Time t would usually be the "time sent" information recorded on the receiver replica clock as received from the SV.

So what is t_k used for? As stated above the ephemeris data define an orbit path defined at t_{oe} time. For times other than t_{oe} we need to use the correction terms to modify the path a bit. Those correction terms and the equations that incorporate them use t_k to refine the SV position estimate. To be absolutely clear, t_k is a computed value from information gathered by the receiver. It is not sent as part of the ephemeris message (Fig. 4.3).

4.3.5 Computing t_k for Any Given "Time Sent"

We can easily determine t_k for the point we wish to calculate the SV position from difference in the "time sent" and t_{oe}:

Let, $t = \text{Tsent},$ then

$$t_k = t - t_{oe}. \tag{4.2}$$

If we find that after doing this calculation that $t_k = 0$ then we are right at the reference position. We would expect computed SV position errors to be the minimum at this time for reasons discussed above.

4.3.6 Comments on the Time Scaling of t_{oe} and t_k

In Sect. 1.10, we added the second counting Dial to our clock model. We commented there about the tic size of this Dial being 1 s and not 1.5 s, which is what the

Parameter	No. of Bits**	Scale Factor (LSB)	Effective Range***	Units	Subframe#	Word#'s	Typical Value
IODE	8			(see text)	2	3	SEE TEXT
C_{rs}	16*	2^{-5}		Meters	3	5	-50.71875
Δn	16*	2^{-43}		# semi-circles/sec	2	4	4.0230e-9
M_0	32*	2^{-31}		# semi-circles	2	4,5	0.1936408
C_{uc}	16*	2^{-29}		radians	2	6	-0.0000024
e	32	2^{-33}	0.03	dimensionless	2	6,7	0.014060
C_{us}	16*	2^{-29}		radians	2	8	0.0000046
$(A)^{1/2}$	32	2^{-19}	604,784	meters$^{1/2}$	2	8,9	5153.6702614
t_{oe}	16	2^{4}		seconds	2	10	338,400
C_{ic}	16*	2^{-29}		radians	3	3	1.1315e-8
Ω_0	32*	2^{-31}		# semi-circles	3	3,4	1.3732476
C_{is}	16*	2^{-29}		radians	3	5	3.7511e-9
i_0	32*	2^{-31}		# semi-circles	3	6	0.9503171
C_{rc}	16*	2^{-5}		meters	3	7	287.53125
ω	32*	2^{-31}		#semi-circles	3	7,8	-2.5648186
$\dot{\Omega}$	24*	2^{-43}		# semi-circles/sec	3	9	-7.9213e-9
IDOT	14*	2^{-43}		# semi-circles/sec	3	10	1.7721e-10
SV CLOCK CORRECTION TERMS (only clock phase bias shown)							
a_{f0}	22*	2^{-31}		Seconds	1	10	-0.00017891

* Parameters so indicated shall be two's complement, with the sign bit (+ or -) occupying the MSB;
** See Figure 20-1 for complete bit allocation in subframe;
*** Unless otherwise indicated in this column, effective range is the maximum range attainable with indicated bit allocation and scale factor.
\# Multiply by π to convert to radians

Fig. 4.3 Ephemeris parameters and SV clock correction terms

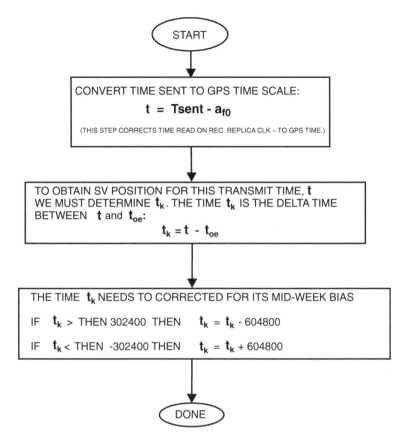

Fig. 4.4 Flow chart for determining t_k

TOW message is scaled to. Well there was a reason and now we have "run into" it. The equations and parameters associated with solving for the SV position expect t_{oe} and t_k to be in the units of seconds not 1.5 s intervals.

This process of getting the "time sent" is now simple once our receiver replica clock(s) are scaled to seconds. For any particular instant of time, we can "snapshot" the receiver's replica clock Dials and compute the time sent by summing each Dials value multiplied by its tic size (see Sect. 1.5).

The calculation of t_k is summarized in the flow chart of Fig. 4.4. Notice that the calculation has one additional complication. The value computed for t_k is "biased" to a mid week value. This results in a simple compare and addition/subtraction operation.

4.4   GPS SV Orbit Description

Figure 4.1 shows a single GPS SV in orbit around the earth. The angles, rates, and times indicated are the primary information needed to solve for the SV position. In addition to this information are second- and third-order terms. These terms are not shown in Fig. 4.1. These other terms are used to correct out perturbations in the orbit from a variety of sources. Figure 4.2 lists the complete set of 16 orbital parameters that are sent down to the user receiver. The user receiver must perform a series of moderately difficult calculations in order to solve for the position of the SV in ECEF coordinates. Figure 4.3 lists the 16 orbital parameters, number of bits, scale factor, variable range, Units, where in the data message it is found and a typical value. For more detailed information about the location and format of the parameters in the data message, see Appendix D.

4.4.1   Solving the Equations for the SV Position and Speed

It is not possible in the confines of this text to cover the details of orbital mechanics and the associated Keplerian equations. Many excellent references exist for the interested reader to pursue if they so desire. A more practical approach will be pursued. Figure 4.5a, b shows the set equations in flow chart form that must be solved in order to get to our goal of the SV position in ECEF coordinates. We can solve for SV coordinates by simply plugging in the parameters and doing the indicated calculations with two exceptions. The first is the calculation of the eccentric anomaly E_k. This term is typically solved for numerically, it is nonlinear equation. The flow chart of Fig. 4.6 is one possible solution to this issue. The other calculated term that requires special attention is the equation for the true anomaly, v_k. Solving for v_k requires the sign of inverse tangent function be properly taken care of. Finally, Fig. 4.7 is C program listing of a program that calculates the SV position using some PRN-2 ephemeris data from Appendix D. Note that the Ephemeris data in Appendix D is not "raw" SV data in that those constants with units of semicircles have been converted to radians.

For the reader interested in computing the approximate orbital speed, the x and y values of the SV in the *orbital plane* (xp, yp) can be used. Compute the position in the orbital plane of the SV at two times separated by 1 s. Then compute the distance between these two points (orbital path is nearly straight line for small delta t). Divide by 1 s and you have the approximate SV speed in meters per second. If the (Xk,Yk,Zk) coordinates are used the result is in error as ECEF axis system is rotating with the earth.

a

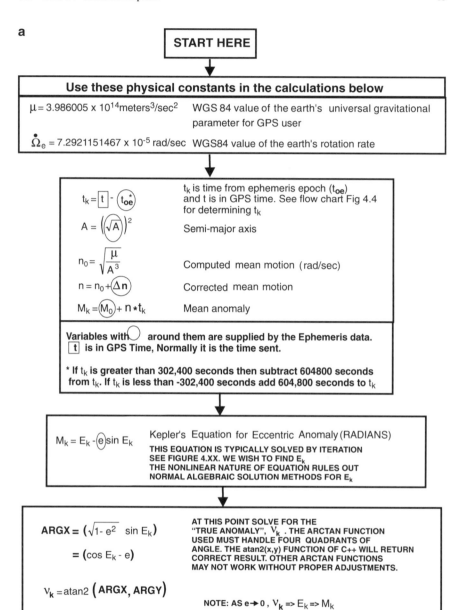

Fig. 4.5 (a) Flowchart for computing SV position in ECEF coordinates from SV ephemeris data and user defined t. (b) Continuation of flowchart for computing SV position in ECEF coordinates from SV ephemeris data and user defined t

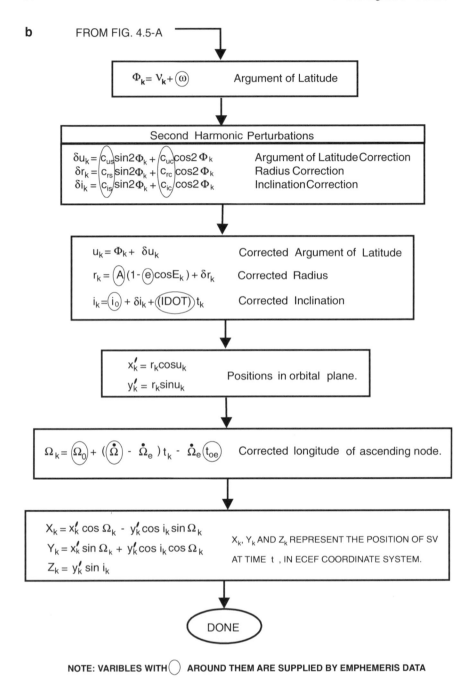

b   FROM FIG. 4.5-A

$\Phi_k = V_k + \textcircled{ω}$ Argument of Latitude

Second Harmonic Perturbations

$\delta u_k = \textcircled{c_{us}} \sin 2\Phi_k + \textcircled{c_{uc}} \cos 2\Phi_k$ Argument of Latitude Correction
$\delta r_k = \textcircled{c_{rs}} \sin 2\Phi_k + \textcircled{c_{rc}} \cos 2\Phi_k$ Radius Correction
$\delta i_k = \textcircled{c_{is}} \sin 2\Phi_k + \textcircled{c_{ic}} \cos 2\Phi_k$ Inclination Correction

$u_k = \Phi_k + \delta u_k$ Corrected Argument of Latitude

$r_k = \textcircled{$A$}(1 - \textcircled{$e$}\cos E_k) + \delta r_k$ Corrected Radius

$i_k = \textcircled{$i_0$} + \delta i_k + (\textcircled{IDOT})t_k$ Corrected Inclination

$x_k' = r_k \cos u_k$
$y_k' = r_k \sin u_k$ Positions in orbital plane.

$\Omega_k = \textcircled{$\Omega_0$} + (\textcircled{$\dot{\Omega}$} - \dot{\Omega}_e)t_k - \dot{\Omega}_e\textcircled{t_{oe}}$ Corrected longitude of ascending node.

$X_k = x_k' \cos \Omega_k - y_k' \cos i_k \sin \Omega_k$
$Y_k = x_k' \sin \Omega_k + y_k' \cos i_k \cos \Omega_k$
$Z_k = y_k' \sin i_k$

X_k, Y_k AND Z_k REPRESENT THE POSITION OF SV
AT TIME t , IN ECEF COORDINATE SYSTEM.

DONE

NOTE: VARIBLES WITH $\bigcirc$ AROUND THEM ARE SUPPLIED BY EMPHEMERIS DATA

Fig. 4.5 (continued)

Fig. 4.6 Flow chart for
determining E_k

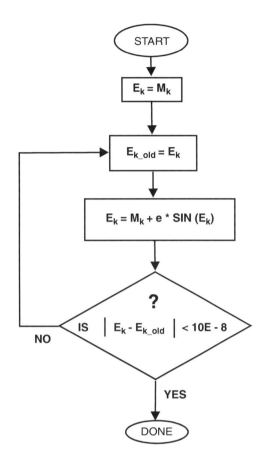

4.4.2   Second- and Third-Order Correction Terms

As previously mentioned, Fig. 4.1 does not show a number of the ephemeris
parameters. The parameters of IDOT, OMEGADOT, and Δn are second-order
terms that primarily account for variations in the orbit due to variations in the
earth's gravity field due to nonspherical and other effects. These effects, if not
compensated for, would lead to large errors (thousands of meters) in computed SV
position once we move away from the reference time, t_{oe}. The remaining
parameters fine-tune the calculations to account for other perturbations to the
orbit. Without the six terms C_{uc}, C_{us}, C_{rc}, C_{rs}, C_{ic} and C_{is}, the error in computed
position of the SV could be in the hundreds of meters. Given our goal of 100 m
accuracy we could possibly ignore these last terms, certainly if the present time is
close to the reference time t_{oe}.

```
/***********************************************************************
FUNCTION satpos()

PURPOSE:
THIS SUBROUTINE CALCULATES THE SATELLITE POSITION IN ECEF COORDINATES FROM SUPPLIED EPHEMERIS
DATA.
```

Mo	- MEAN ANOMALY
Dn	- MEAN MOTION DIFFERENCE
ecen	- ORBITAL INITIAL ECCENTRICITY
SQRA	- SQUARE ROOT OF SEMIMAJOR AXIS
Omega0	- LONGITUDE OF ASCENDING NODE
Io	- INCLINATION ANGLE
w	- ARGUMENT OF PERIGEE
Omegadot	- RATE OF RIGHT ASCENSION
IDOT	- RATE OF INCLINATION ANGLE
Cuc	- ARGUMENT OF LATITUDE COSINE CORRECTION TERM
Cus	- ARGUMENT OF LATITUDE SINE CORRECTION TERM
Crc	- RADIUS COSINE CORRECTION TERM
Crs	- RADIUS SINE CORRECTION TERM
Cic	- INCLINATION COSINE CORRECTION TERM
Cis	- INCLINATION SINE CORRECTION TERM
toe	- EPHEMERIS REFERNCE TIME
t	- TIME THAT SV POSITION DESIRED (GPS TIME)
tk	- DELTA TIME FROM t TO toe
Omegae	- EARTHS ROTATION RATE
mu	- GRAVATIONAL CONSTANT

```
***********************************************************************/

double satpos(t,Xx,Yy,Zz,xpl,ypl)
double t, *Xx, *Yy,*Zz, *xpl, *ypl;
{
    double Mk,Ek,diff,vk,aol,delr,delal,delinc,rk,inc;
    double Ia,xp,yp,tk,A,mu,Omegae;
    double no,Xk,Yk,Zk;

// define the 16 Ephemeris variables
double Mo,Dn,ecen,SQRA,Omega0,Io,w,Omegadot,IDOT,Cuc,Cus;
double Crc,Crs,Cic,Cis,toe,ARGX,ARGY,EkOLD;

        mu=3.986005E14; Omegae=7.2921151467E-5;

//  initialize Ephemeris Data. All angles must be in Radians
        Mo =2.23108606228;
        Dn = 5.08342603081E-9;
        ecen = 0.0228992204648;
        SQRA = 5153.55303764;
        Omega0 = 5.97931689194;
        Io = 0.932615670775;
        w = -1.82126037919;

Omegadot = -7.94033074665E-9;
        IDOT = -9.27895793441E-10;
        Cuc = -3.05287539959E-6;
        Cus = 1.02911144495E-5;
        Crc = 155.75;
        Crs = -52.78125;
        Cic = 2.73808836973E-7;
        Cis = -3.55765223503E-7;
        toe = 453600.0;

    // Compute time difference from ephemeris ref time.
    tk = t - toe;
    if (tk>302400.0) tk = tk-604800.0;
                else if (tk<-302400.0) tk=tk + 604800.0;

    // Compute mean Motion, no
    A= SQRA*SQRA;
    no = sqrt(mu/(A*A*A));
```

Fig. 4.7 SV position calculations

```
                    // compute Mean Anomaly
                    Mk = Mo + tk*( no + Dn);

                    // Solve for Eccentric Anomaly by iteration
                    Ek = Mk;
                    do
          {
                    EkOLD=Ek;
                    Ek = Mk + ecen*sin(Ek);
                    diff= Ek - EkOLD;
                    }while (fabs(diff) > 1.0e-8 );

//          Compute the True Anomaly, vk (Angle from Perigree)
                    ARGX= sqrt(1.00-ecen*ecen) * sin(Ek);
                    ARGY= cos(Ek)-ecen;
                    vk=atan2(ARGX,ARGY);

          //Calculate aol,The Argument of Lat. Of the SV
                    aol= vk + w;

//          Calculate the second harmonic perturbations of the orbit
          delr  = Crc*cos(2.0*aol)+ Crs*sin(2.0*aol);
          delal = Cuc*cos(2.0*aol)+ Cus*sin(2.0*aol);
          delinc= Cic*cos(2.0*aol)+ Cis*sin(2.0*aol);

          // Corrected argument of latitude
          aol=aol+delal;

          // rk is the radius of SV at time t
          rk=(A)*(1.00- ecen*cos(Ek))+delr;

// Corrected Inclination Angle
          inc= Io + delinc + IDOT*tk;

                    //calculate positions in orbital plane
                    xp=rk*cos(aol);    *xpl=xp;
                    yp=rk*sin(aol);    *ypl=yp;

                    //LA IS THE CORRECTED LONGITUDE OF THE ASCENDING NODE
                    la=Omega0 + (Omegadot - Omegae)*tk - Omegae*toe;

                    // calculate  X, Y, Z of SV
                    Xk=xp*cos(la)-yp*cos(inc)*sin(la);

                    Yk=xp*sin(la)+yp*cos(inc)*cos(la);

                    Zk=yp*sin(inc);

//          printf(" X position %f", Xk);
//          printf(" Y position %f", Yk);
//          printf(" Z position %f", Zk);

                    *Xx=Xk;  *Yy=Yk; *Zz=Zk;


          return(0);
          }// end satpos()
```

Fig. 4.7 (continued)

4.4.3 Some Comments on the Ephemeris Data and Solving for the SV Position

The equations for the SV position and the ephemeris data that feeds them form a fairly complex task. There are a number of observations that may help the new comer with these calculations:

- Before the raw Ephemeris data can be used it must be scaled. The Scaling terms can be found in Fig. 4.3. Each raw binary parameter must be multiplied by these terms before use in the equation set for the SV position.
- Seven of the parameters have an unusual unit of semicircles. Multiply by π in addition to the scale factor to convert these terms to radians.
- The solution for E_k and V_k anomalies (which are angles) is complicated by an iteration task and a inverse tangent. Note that as the eccentricity term goes to zero (if $e = 0$ orbit is a perfect circle) we can make the statement $V_k = E_k = M_k$. Since e is normally small V_k and E_k are very close in value to M_k.
- The correction terms IDOT, Omegadot, Δn, and the six harmonic correction terms are small. For a pretty good estimate of the SV position these terms can be set to zero and solution pursued. For the harmonic terms this means that all the δ terms are set to zero.
- If t_k is small (<500) the terms it multiplies with in corrected longitude of ascending node calculation and the corrected inclination angle equations can be estimated to be zero for a rough fix of the SV position.

4.4.4 Other SV Orbit Information, Almanac Data

As we mentioned earlier each SV sends down as part of its data stream a rough estimate of all the other SV orbits. This is called Almanac Data. When the user receiver is getting its first position fix this information can be used to help determine which SV are possibly in view and estimate there approximate positions. It is not necessary to use this information assuming one has four SV in view with good ephemeris data for their orbits available. Rather this information is typically used to speed up the first fix and for knowing system health. It is not the purpose of this text to thoroughly explore the Almanac data and its uses.

4.5 Age/Issue of Data Terms

The data stream contains a number of terms that give the user receiver information about how old the Ephemeris and SV clock error data are or if it has changed in the middle of a data frame. These terms are *IODE*, *IODC*, *AODO*, and *AODE*.

The *I*ssue *O*f *D*ata *E*phemeris (IODE) lets the user know if the data have changed in the middle of a data frame. The clock correction data have a related term *I*ssue of *C*lock *D*ata (IOCD). The term age of data ephemeris (AODE) lets the user know how old the ephemeris data are without knowledge of current GPS time. The AODO term is information about the age of the almanac data.

For our purposes, the IODE term is the most important. We are not that concerned if the ephemeris data are a bit old but we do want to know if it is valid for data frame we are receiving. The details of using this term are found in the ICD-200°C document.

4.6 t_{oc} SV Clock Reference Time

The clock error terms a_{f0}, a_{f1}, and a_{f2} have a reference time just like the ephemeris data do. The time t_{oc} is sent in subframe 1 with clock error terms and is referenced to the GPS time scale. We did not mention this term until now, as its use is optional for our goal obtaining position to ±100 m and one more "time variable" increases the confusion. If time t is equal to t_{oc} then the error of the SV clock is just the term a_{f0}. As GPS time moves away from the t_{oc} point, the SV clock will drift at the drift rate given by a_{f1}. This drift rate is small. Inspection of the ephemeris data in Appendix D shows that the SV clock drift rate, a_{f1}, is typically below 3.0×10^{-12} seconds per second. In 1 h this drift rate would accumulate to $(3.0 \times 10^{-12}) \times 3{,}600 \sim 10$ ns. This would be the unaccounted SV clock error after 1 h if we used only the a_{f0} term to correct the SV time. It is approximately 3 m of path delay error per hour. The point of doing this calculation is to show that we can safely ignore the higher-order clock corrections as the error introduced by just using static model is small. The clock error terms are typically updated with the ephemeris data, which is approximately on an hourly basis. The ICD-200°C document has a very complete (and complex) SV clock correction algorithm for those wishing to pursue more accuracy.

4.7 Summary

In this chapter, we have seen how to solve for the SV position at the time sent moment. The structure and use of the ephemeris data was explored. We also investigated some of the details of solving the SV position problem from received ephemeris data. We covered the SV clock error model and the correction terms as sent in the data from the SV to user. We now have all the information we need to solve for the user position. Solving the equations for user position is a task in itself and is covered in the next chapter.

Chapter 5
Solving for User Position

In this chapter, we will investigate solving for the User's position on the earth's surface. A brief derivation of a linear set of equations is presented. The solution will determine the user's position in ECEF coordinates and the user's clock bias. A flow chart is presented implementing the solution described. Then a conversion is done to arrive at user position in Latitude, Longitude, and Altitude.

5.1 Iteration Versus Direct Solution

In Chap. 2 we introduced the equations that if solved will produce the user position and the receiver's clock bias. These equations are shown in Fig. 2.7. Due to the square root, squaring operations, and simultaneous nature of these equations their solution is not trivial. One method would be to "guess" or estimate a user position and user clock bias, plug in the numbers and see if it "works". This is a very crude approach and without refinement would not produce a term that indicates the direction to move the "estimates" in order to improve the solution.

5.2 Linear Approximation

A very common method used to solve the equations of Fig. 2.7 is to use a mathematical approach that will "linearize" the equations. This method also uses an assumed user position and assumed user clock error. The assumed user position and clock error is referred to as the *nominal* values. In addition, *delta* terms are introduced that when added to the nominal estimate of user position and clock error produces the *true* user position and clock error. In the following solution, we will assume that four SV positions are available. Solutions that use more than four SVs will not be pursued here.

D. Doberstein, *Fundamentals of GPS Receivers: A Hardware Approach*,
DOI 10.1007/978-1-4614-0409-5_5, © Springer Science+Business Media, LLC 2012

Specifically. We will drop the (x_u, y_u, z_u) notation we used earlier for user position and let (x, y, z) represent the *true* user position. As before, we will let T_{bias} be the *true* user clock error.

Let the nominal point be defined by (x_n, y_n, z_n) and the nominal clock error be represented by T_{bn}.

We also need to define the relationship between the true user position/clock error, nominal user position/clock error and the difference between them. These ideas can be expressed as:

$$x = x_n + \Delta x \tag{5.1}$$

$$y = y_n + \Delta y \tag{5.2}$$

$$z = z_n + \Delta z \tag{5.3}$$

$$T_{bias} = T_{bn} + \Delta T_b \tag{5.4}$$

We can see from these equations that if we can force the delta terms to zero, the nominal or estimate for position and clock error will equal the *true* user position and *true* clock error. The iteration method we will introduce does this by computing the delta terms. It then examines the delta terms. If they are small enough, a solution is declared. If the delta terms are not small enough, then adding the delta terms to the present nominal point creates a new nominal point. This new nominal point is then processed and a new set of delta terms is calculated. It is important to emphasize that the equations do not solve directly for the user position/clock error but instead for the deltas involved. This process is repeated or *iterated* until the delta terms are small enough.

5.3 The Pseudo Range Equations for Four SV

Before we can proceed we need to revisit some of the equations from Fig. 2.7. In the figure, we showed the true range to each SV from the user as R_i. If we rewrite these equations with the user clock error term added to them, we have the Pseudo Ranges to each SV. We use (x, y, z) to represent the true user position. The set of equations expressing the Pseudo-Range to each SV is:

$$PR_1 = [(x - x_{sv-1})^2 + (y - y_{sv-1})^2 + (z - z_{sv-1})^2]^{1/2} + T_{bias} \times C \tag{5.5}$$

$$PR_2 = [(x - x_{sv-2})^2 + (y - y_{sv-2})^2 + (z - z_{sv-2})^2]^{1/2} + T_{bias} \times C \tag{5.6}$$

$$PR_3 = [(x - x_{sv-3})^2 + (y - y_{sv-3})^2 + (z - z_{sv-3})^2]^{1/2} + T_{bias} \times C \tag{5.7}$$

$$PR_4 = \left[(x - x_{sv-4})^2 + (y - y_{sv-4})^2 + (z - z_{sv-4})^2\right]^{1/2} + T_{bias} \times C \qquad (5.8)$$

where PR_i is the Pseudo Range to each SV.

5.4 Forming the Nominal Pseudo-Range

We can use the nominal user position and nominal clock error to form a nominal Pseudo-Range PR_n. PR_n is the range (distance) from the user to each SV calculated using the nominal point (x_n, y_n, z_n) and the known SV positions with assumed nominal clock error included as a distance (multiply T_{bn} by C). We again assume four SVs; therefore there are four PR_n's:

$$PR_{n1} = \left[(x_n - x_{sv-1})^2 + (y_n - y_{sv-1})^2 + (z_n - z_{sv-1})^2\right]^{1/2} + T_{bn} \times C \qquad (5.9)$$

$$PR_{n2} = \left[(x_n - x_{sv-2})^2 + (y_n - y_{sv-2})^2 + (z_n - z_{sv-2})^2\right]^{1/2} + T_{bn} \times C \qquad (5.10)$$

$$PR_{n3} = \left[(x_n - x_{sv-3})^2 + (y_n - y_{sv-3})^2 + (z_n - z_{sv-3})^2\right]^{1/2} + T_{bn} \times C \qquad (5.11)$$

$$PR_{n4} = \left[(x_n - x_{sv-4})^2 + (y_n - y_{sv-4})^2 + (z_n - z_{sv-4})^2\right]^{1/2} + T_{bn} \times C \qquad (5.12)$$

where C is speed of light

5.5 Forming the Estimate of the Pseudo-Range to Each SV

We can form an estimate of the Pseudo-Ranges PR_i by using our calculated nominal Pseudo-Ranges PR_{ni}. We can write the corresponding true range PR_i to nominal Pseudo-Range PR_n relation (for each SV to User Range) just like we did with the user position and nominal position as a nominal Pseudo-Range added to a delta error term:

$$PR_1 = PR_{n1} + \Delta PR_1 \qquad (5.13)$$

$$PR_2 = PR_{n2} + \Delta PR_2 \qquad (5.14)$$

$$PR_3 = PR_{n3} + \Delta PR_3 \qquad (5.15)$$

$$PR_4 = PR_{n4} + \Delta PR_4 \qquad (5.16)$$

Where the term ΔPR_i is the error in the nominal Pseudo-Range estimate with respect to the true Pseudo Range.

5.6   Resulting Linear Equation Set

Equations (5.1)–(5.4) and (5.13)–(5.16) can be combined with (5.5)–(5.8) to form a set of four simultaneous equations. We will not cover the details of this operation but will state the result:

$$\Delta PR_1 = \alpha_{11}\Delta x + \alpha_{12}\Delta y + \alpha_{13}\Delta z + C \times \Delta T_b \tag{5.17}$$

$$\Delta PR_2 = \alpha_{21}\Delta x + \alpha_{22}\Delta y + \alpha_{23}\Delta z + C \times \Delta T_b \tag{5.18}$$

$$\Delta PR_3 = \alpha_{31}\Delta x + \alpha_{32}\Delta y + \alpha_{33}\Delta z + C \times \Delta T_b \tag{5.19}$$

$$\Delta PR_4 = \alpha_{41}\Delta x + \alpha_{42}\Delta y + \alpha_{43}\Delta z + C \times \Delta T_b \tag{5.20}$$

where the α_{ij} coefficients are defined as;

$$\alpha_{i1} = [x_n - x_{sv-i}]/[PR_{ni} - T_{bn} \times C] \tag{5.21}$$

$$\alpha_{i2} = [y_n - y_{sv-i}]/[PR_{ni} - T_{bn} \times C] \tag{5.22}$$

$$\alpha_{i3} = [z_n - z_{sv-i}]/[PR_{ni} - T_{bn} \times C] \tag{5.23}$$

where $i = 1, 2, 3$ or 4

The set of simultaneous equations shown above is now linear in the unknowns Δx, Δy, Δz, and ΔT_b. With the exception of the ΔPR_i terms, the other terms in the (5.17)–(5.20) are from measured data, known or in the case of the nominal values "guessed" (initially). But before we can finish our solution, we need to find a way to compute the term ΔPR_i using known information or measured data.

5.7   Estimating the ΔPR_i Term

Equations (5.17)–(5.20) contain the Pseudo-Range term ΔPR_i. To investigate these terms, let us rearrange and compact the equations (5.13)–(5.16) as:

$$\Delta PR_i = PR_i - PR_{ni} \tag{5.24}$$

Now we can see that each ΔPR_i term is the difference in the *true* Pseudo-Range and the nominal Pseudo-Range. This follows from our statements about the user position (x, y, z) as being the *true* user position and T_{bias} being the *true* user clock error. Using these values in (5.5) through (5.8) means that each Pseudo-Range PR_i is the *true* one and not an estimate. This creates a problem, as we do not have the true values of user position or user clock error. We do, however, have an estimate of the true Pseudo-Range and that is the *measured* path delay. Specifically, we can estimate the ΔPR_i's as the difference in the measured path delay and the nominal Pseudo-Range:

$$\Delta PR_1 \approx \Delta t_1 \times C - PR_{n1} \tag{5.25}$$

$$\Delta PR_2 \approx \Delta t_2 \times C - PR_{n2} \tag{5.26}$$

$$\Delta PR_3 \approx \Delta t_3 \times C - PR_{n3} \tag{5.27}$$

$$\Delta PR_4 \approx \Delta t_4 \times C - PR_{n4} \tag{5.28}$$

where each Δt_i is defined by;

$$\Delta t_1 = Trec - Tsent_{sv-1} + Terr_{sv-1} - Tatm_{sv-1} \tag{5.29}$$

$$\Delta t_2 = Trec - Tsent_{sv-2} + Terr_{sv-2} - Tatm_{sv-2} \tag{5.30}$$

$$\Delta t_3 = Trec - Tsent_{sv-3} + Terr_{sv-3} - Tatm_{sv-3} \tag{5.31}$$

$$\Delta t_4 = Trec - Tsent_{sv-4} + Terr_{sv-4} - Tatm_{sv-4} \tag{5.32}$$

Note the definition used here for Δt_i is different than used in Fig. 2.7. Specifically the T_{bias} term is omitted. We have again assumed that the "receive" time is the same for all four path-delay measurements. The nominal Pseudo-Ranges are calculated from the nominal values and the measured SV positions. The *true* Pseudo-Ranges are calculated from the measured path delay (Trec–Tsent) and knowledge of SV clock errors and atmospheric Delays. With this definition of ΔPR_i, we can proceed with the solution.

5.8   Matrix Form of Solution

The entire set of equations presented in (5.17)–(5.20) can be written in matrix format as:

$$\begin{bmatrix} \alpha_{11} & \alpha_{12} & \alpha_{13} & C \\ \alpha_{21} & \alpha_{22} & \alpha_{23} & C \\ \alpha_{31} & \alpha_{32} & \alpha_{33} & C \\ \alpha_{41} & \alpha_{42} & \alpha_{43} & C \end{bmatrix} \times \begin{bmatrix} \Delta x \\ \Delta y \\ \Delta z \\ \Delta T_h \end{bmatrix} = \begin{bmatrix} \Delta PR_1 \\ \Delta PR_2 \\ \Delta PR_3 \\ \Delta PR_4 \end{bmatrix} \tag{5.33}$$

The terms are still defined by equations (5.21)–(5.23). Assuming the 4×4 matrix is invertible, (5.32) can be rewritten as:

$$
\begin{bmatrix} \Delta x \\ \Delta y \\ \Delta z \\ \Delta T_b \end{bmatrix} = \begin{bmatrix} \alpha_{11} & \alpha_{12} & \alpha_{13} & C \\ \alpha_{21} & \alpha_{22} & \alpha_{23} & C \\ \alpha_{31} & \alpha_{32} & \alpha_{33} & C \\ \alpha_{41} & \alpha_{42} & \alpha_{43} & C \end{bmatrix}^{-1} \times \begin{bmatrix} \Delta PR_1 \\ \Delta PR_2 \\ \Delta PR_3 \\ \Delta PR_4 \end{bmatrix}
\tag{5.34}
$$

Equation (5.34) demonstrates the final solution to finding the user position and clock bias. By solving for the delta terms on the left side of the equation we can test them to see if they are "small enough". If these terms are still too large, we form new nominal values by adding these delta terms to our previous nominal values and redo the calculation as indicated on the right side of the equation.

All the terms on the right hand side are known. In particular, the Assumed Nominal user position (x_n, y_n, z_n), Assumed Nominal User Clock Bias (T_{bn}), the four SV positions from the ephemeris (x_{sv}, y_{sv}, z_{sv}), SV Clock Errors $(Terr_{sv})$, Atmospheric delay $(Tatm_{sv})$ from models and the measured path delay $(Trec–Tsent_{sv})$.

5.9   Flowchart, C Code Program, Assumed Initial Position/User Clock Bias

Figure 5.1a–c shows a flow chart of the steps needed to solve the equations presented here for user position and user clock bias.

The assumed nominal point can be taken as the center of the earth, (0,0,0). The assumed nominal clock bias can be taken as zero also. Once the computer starts iterating, it takes remarkably few "loops" to converge on the solution or more precisely the point when the user position error is less than 30 m or so. Typically, it takes less than ten loops to achieve the solution. Usually if it takes more than ten loops, a problem in the calculations is indicated, and an error is declared. This could be due to geometry problems (see below) or problems with the data provided to the software routines.

5.10   Testing for Solution

A common way to determine if the delta position terms have decreased to point where a solution can be declared is to square root the sum of the squares:

$$
\text{Error Magnitude} = \text{SQRT} \left[\Delta x^2 + \Delta y^2 + \Delta z^2 \right]
\tag{5.35}
$$

This allows a simple test to be performed which will reflect all three distance error terms. For our purposes, when the error magnitude is less than 100 m we declare a solution has been obtained.

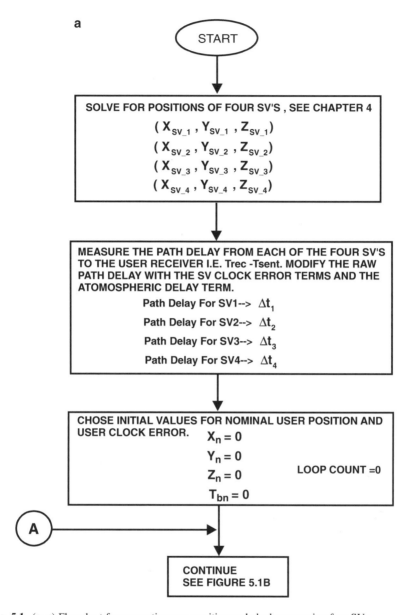

Fig. 5.1 (**a–c**) Flowchart for computing user position and clock error using four SVs

b

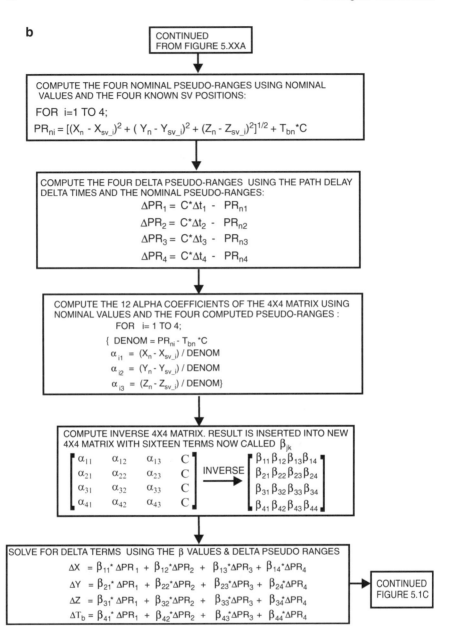

Fig. 5.1 (continued)

c

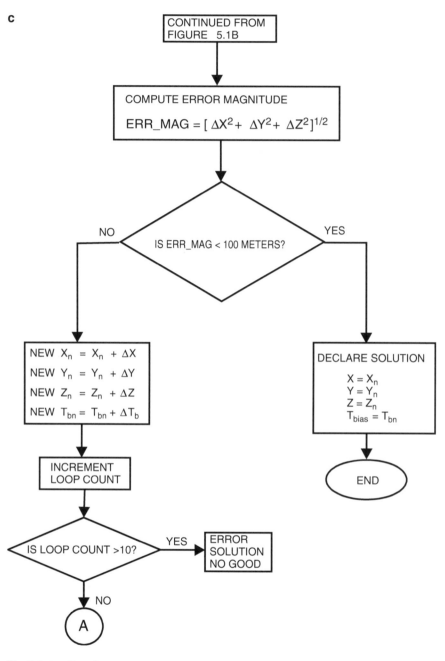

Fig. 5.1 (continued)

5.11   Geometry Considerations

The geometry of the SV used in the computations above can lead to a situation where the loop does not converge on a solution. In other words, the calculations "Blow up". This problem occurs when the four SVs are all in or very near to a common plane. In this scenario, we have lost 3-D triangulation and are attempting to solve the 3-D problem of user position with 2-D information. We will not go into the solutions to this issue here. In general, there are usually more than four SVs in view and a choice of which ones to use in the calculations will alleviate this issue.

5.12   Converting User Position to Lat/Long/Altitude from ECEF Coordinates (Spherical Earth)

Figure 5.2 shows the geometry involved in converting from user position (x, y, z) in ECEF coordinates to user position in latitude, longitude, and altitude. For clarity, the user's altitude is greatly exaggerated in Fig. 5.2. The calculations reflect an assumed spherical earth.

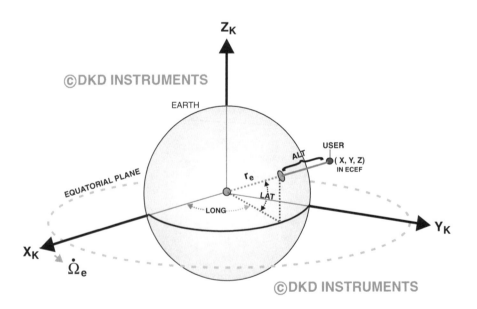

NOTE: ALTITUDE OF USER EXAGGERATED FOR CLARITY

Fig. 5.2 Converting user position in ECEF coordinates to Lat./Long./Alt

The equations for the conversion are:

User altitude $= r - r_e$ Where r_e is the average radius of earth and

$$r = [x^2 + y^2 + z^2]^{1/2} \quad r_e \approx 6,368 \text{ km} \tag{5.36}$$

$$\text{User latitude } (L_{gc}) = \tan^{-1}[z/\zeta] \text{ where } \zeta = [x^2 + y^2]^{1/2} \tag{5.37}$$

$$\text{User Longitude} = \tan^{-1}[y/x] \tag{5.38}$$

5.13 Corrections for Non-Spherical Earth

The equations above assumed a perfect spherical earth. The earth is not perfectly spherical but instead slightly bulged at the equator. The approximate corrections for this effect are shown below. The effect of the earth's bulge induces errors (primarily) in the computed user latitude and altitude. For this reason, the calculation for the longitude is unchanged. The equation for corrected latitude must be solved by iteration very much like we saw with the *true anomaly* in solving for SV position; see Fig. 4.5.

When the correction is made to the user's latitude, it is now a Geocentric latitude (L_{gd} see below). The latitude solved for in (5.37) is the Geodetic latitude commonly used on maps

$$\text{(Corrected User Lat.)} L_{gd} \approx L_{gc} + e_p \sin(2L_{gd}) e_p$$
$$= 0.0033528106645 (\text{ecen. earth}) \tag{5.39}$$

$$\text{(Corrected Radius earth)} r_0 \approx a_e (1 - e_p \sin^2 L_{gd}) a_e$$
$$\approx 6,378,137 \text{ m (semi major axis of earth)} \tag{5.40}$$

$$\text{Corrected user altitude} \approx r - r_0 \tag{5.41}$$

5.14 Summary

This chapter introduced the linear method for solving the four simultaneous equations associated with finding the users position and clock error. The position solution was detailed in terms of a flow chart. The conversion of the user position to latitude, longitude, and altitude was also discussed with details. In addition, the corrections to user position for non-spherical earth was presented.

Part II

Chapter 6
GPS Receiver Hardware Fundamentals

This chapter presents a broad overview of the hardware behind an analog implementation of a GPS receiver. The hardware's goal is to get the SV data, replicate the SV clock, properly "set" the replica clock and make the "time sent" and "time received" measurements to achieve the goal of computing user position. The functions of Doppler Scan/Track, C/A Code Scan/Track, and data clock recovery are introduced and discussed here, as they are not typical to conventional narrow band receivers. Chapter 7 will focus on the details of the hardware subsystems presented in this chapter.

6.1 Analog Versus Digital GPS Receivers

All early GPS receivers were primarily analog at the correlator and usually analog subsequent to this point. Also many early designs were *synchronous* in operation. These designs allowed the recovery of the code clock by tracking the Doppler. This often resulted in restrictions to the frequency plan of the receiver. *Coherent* designs use a common reference that the LOs are derived from but the requirement for total coherence is dropped. The hardware described in this chapter and Chap. 7 is *almost* coherent but it has areas that are not coherent.

A valid question is why would one want to study an older analog design? Many may feel that analog is "dead" and its a waste of time pursuing an analog design. Its the author's view that these older designs are still a valid teaching tool. The approach we will pursue in the "all analog receiver" for C/A code tracking is the Tau-Dither method. A variation of this method is used in the Zarlink chip set, which is an All-Digital receiver (after its analog down conversion). The differences in operation between an all digital version and the analog version of Dither-based code tracking are fairly minor. The point is Tau-Dither is the "Grand pappy" of this digital "Dither" approach. Taking the time to understand the Tau-Dither approach is not wasted. Once we understand the older method, the bridge to the Digital implementations is easily crossed.

D. Doberstein, *Fundamentals of GPS Receivers: A Hardware Approach*,
DOI 10.1007/978-1-4614-0409-5_6, © Springer Science+Business Media, LLC 2012

In short in this chapter it is not the authors intent to present a "state-of-the-art" receiver but rather cover in detail a "simple" analog design uses digital methods in conjunction with analog methods to achieve code lock. By doing this it is hoped that the student of GPS receivers can learn how the old influenced the new. It also will illustrate some of the many ways the common GPS receiver design problems can be solved and the commonality of past and (more) modern solutions.

6.2   Five Fundamental Steps in the GPS Receiver Hardware

The GPS receiver hardware must perform a number of steps to achieve the goal of the demodulated data and the path delay measurements. These steps can be broken down into five fundamental processes: RF Down-conversion, Signal Acquisition, Data Demodulation, Data Clock Recovery, SV Clock Replication, and the "Time Sent-Received information." The last step of forming the time difference typically requires a computer.

6.2.1   Receive RF and Convert to a Lower IF

Most all-current GPS receivers employ some sort of down conversion of the carrier to a lower IF frequency for processing. The RF carrier is to high for direct processing or sampling. A review of current designs shows typically a two IF system. The first IF is usually somewhere from 30 to 200 MHz. The second IF is normally where all the bulk of the processing occurs or it could be the point at which digitization occurs.

6.2.2   Signal Acquisition

Once converted to an IF frequency, the receiver now must "acquire" the signal. The process of Signal Acquisition involves searching the entire range of signal Doppler while simultaneously looking for C/A code correlation. Once the acquisition and correlation in process is accomplished then we can demodulate the data and recover the 50 Hz data clock. For readers not familiar with correlation as used in GPS receivers, see Appendix A. The correlation process will automatically "set" the 0.977 μs and 1 ms Dials to the correct values in the SV replica clock(s).

6.2.3   Data Demodulation and Data Clock Recovery

The 50 Hz data can be recovered by many different methods. The most popular is the Costas Loop method. The method used here is based on quadrature demodulators. Once the data demodulation is done we can move on to recovery of the Data clock. Even though the Data clock frequency is a byproduct of the C/A code correlation process (dividing the 1 kHz epoch by 20), the proper phase of the data clock must still be recovered. By using the timing edges of the 50 Hz data stream (from the SV), the phase of the data clock can be correctly recovered. This will correctly "set" the 20 ms Dial of the received SV replica clock.

6.2.4   Properly Set the 1 s and Second-Counting Dials

The 1 s dial and the Second-Counting Dials must be "set" by using the timing embedded in the 50 Hz data stream. Specifically, the 1-s Dial can be correctly set by using the leading edge of any subframe. Reading the TOW count available at the beginning of every subframe allows the Second-Counting Dial to be properly set (TOW count references the count to Bit1 of the next subframe).

6.2.5   Measure Tsent and Trec

With the recovered SV replica clock for each channel, the receiver is ready to measure Trec and Tsent by recording the "state" of its own reference clock and the SV replica clocks. The method of recording the time will be the SNAP_SHOT method where SNAP_SHOT instant is based on receiver's clock at 1/10 s intervals. The exact implementation of the receiver's reconstructed SV clocks will determine the resolution of the measurement. From this and correction terms the path delay is calculated. Once we have the satellite-supplied ephemeris data and our computed path delay, we can calculate user position.

6.3   Block Diagram of Shared Signal Processing for a Single-Channel Receiver

Figure 6.1 shows a block diagram of the single-channel GPS receiver as implemented in DKD Instruments Model GPS100SC. We will now discuss each element of Fig. 6.1 in some detail.

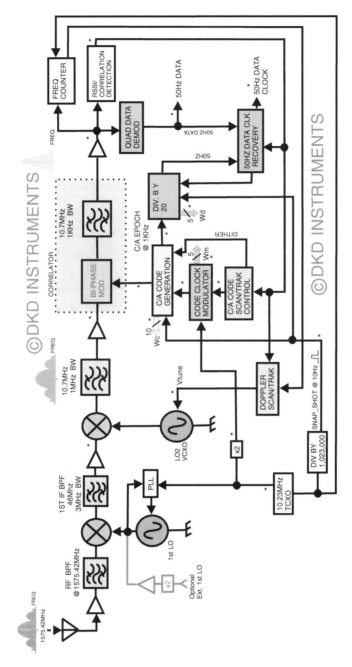

Wd is 5 bits (20 is max value)@1msec/bit, Wc is10 bits @ 0.977517usec/bit, Wm is 5 bits(20 is max value)@ 48.8759nsec/bit

Estimate of Received SV time @ SNAP_SHOT instant is = Zcounts(sec) + N*20msec + Wd*1msec + Wc*0.977517usec + Wm*48.8759nsec - T sv_err

where N is bit number counting from Data Bit 1 of Frame 1. Note: (*) Indicates a test point accessible to user.

Fig. 6.1 Block diagram of GPS100SC receiver

6.3.1   Antenna

The GPS antenna serves as an interface between the GPS signal traveling through free space and the electronics of the receiver. The antenna is a gain element, frequency selective element, and a "matching" device between free space impedance and the impedance of the circuit it is connected to. In addition, the signal from the GPS satellite is Right Hand Circularly Polarized. An optimum GPS antenna should also be circularly polarized, though a linear polarized antenna can be used.

Typically the user receiver does not know where in the sky the SVs are until his position is known. Therefore, a user antenna with low gain or broad beam-width is desired. The optimum gain pattern for the antenna is slightly less than half hemisphere. The "slightly less than" requirement coming from the fact that an SV close to the horizon is often blocked by terrain, etc. The power specified in Fig. 3.5 is for a gain of 3 dBi with a linear polarized antenna. By using a 5 dBi antenna that is circularly polarized, an increase in signal power of approximately +5 dB can be obtained for elevation angles above ~ 20° as compared to the power levels of Fig. 3.5

6.3.2   Pre-amp

The received signal coming out of the antenna is very low in power. So a pre-amp is used to boost signal power to point where subsequent loss will not degrade the SNR as measured at the antenna output. Of course every amplifier has internal noise so there will be SNR loss from the pre-amp. But if the pre-amp gain is high enough and subsequent stages do their part to maintain SNR, then the overall "system" noise figure is set by the Noise figure of the pre-amp.

Currently, pre-amps with noise figures below 1 dB are available. Many commercial GPS antennas come with integrated pre-amps. The noise figures typically range from 1 to 2 dB.

6.3.3   Bandpass and Mixer Stages Up to Second IF

After the pre-amp, a bandpass filter at 1,575.42 MHz is encountered. This filter is used to reject the image frequency of the first mixer. It also provides some protection for out of band signals that may pose a "overload" or saturation threat to the receiver. These filters can be made from many technologies including SAW, Ceramic, Helical, Inter-digital, and microwave strip line. An inter-digital design is presented in Chap. 7.

Following the bandpass at 1,575.42 MHz, the first mixer is encountered. Here, the GPS signal is down-converted to the first IF frequency which is common to all channels.

The signal images for the next conversion are attenuated by this bandpass. The first LO which is mixed with received signal at 1,575.42 MHz is a phase-locked VCO. It is locked to the 20.46 MHz reference.

6.4   Overview of Second IF Processing

It is at the second IF that things start to happen. The tasks of Doppler Scan/Track and C/A Code Scan Track and Data Demodulation are done by signal processing operations on the second IF signal. It is at the second mixer where the Doppler present on the SV signal will be tracked out. We will now discuss each of these functions in more detail.

6.4.1   The Doppler Scan/Track Subsystem

We need to fill in some more details about Doppler and its effects on receiver implementation. As we know from our work in Chap. 3 the GPS signal as received by the user equipment has Doppler on it, up to ± 5 kHz. We also know from that brief discussion that we cannot just make the receiver's bandwidth large enough to pass the full frequency range of the correlated signal. Now that we are examining the hardware we can be more specific about the details of the Doppler problem.

The received GPS signal power is very low. This low power will result in a low SNR, if we use a "large" filter bandwidth. If the subsequent circuits following the second conversion were wide enough to pass the (Correlated) signal and its full range of Doppler they would have to have a bandwidth of approximately 12 kHz. This would be the combined Doppler offset and LO1 error with a ± 0.5 PPM reference. If we compute the SNR of our correlated signal with a 12 kHz bandwidth it is around 5 dB. This will not allow the C/A code-tracking loop to do its job and without correlation we have no signal. Therefore, we must limit the bandwidth of the second IF to approximately 1 kHz or narrower and the primary reason we must do this is the low signal power. Stating the problem a different way is to realize that if the GPS SV transmitted with approximately 10 dB more power we could use a wide enough IF bandwidth that would ensure the correlated signal is in its pass-band.

Now we have a problem. If the VCXO is not set exactly right the signal we want it will not be in our 1 kHz wide second IF filter. To address this problem, we need a method to search for the Correlated signal over the full Doppler range plus our first LO uncertainty due to reference frequency error. Once the signal is found, we must also dynamically track it as it is changing. The offset of LO1 can be estimated from assuming a ± 0.5 PPM reference, this results in an uncertainty in LO1 frequency of ± 770 Hz.

6.4.2   Second Mixer, LO2, and Doppler Scan/Track

These blocks work together to perform the Doppler Scan and Track functions. The frequency of LO2 (a VCXO) is varied by the Doppler Scan/Track block, which controls the Doppler Scan and Track functions. In Scan mode, a sweep or ramp voltage is fed to LO2. This is a slow moving signal that allows the receiver to examine all the possible Doppler offsets for the presence of the GPS signal. As the LO2 sweeps the receiver is also scanning the C/A code, see below. When C/A code correlation is detected the sweeping of LO2 is halted and the Track function activated. In track an error signal (provided by a frequency counter) tells the control circuit which way to move LO2 in order to keep the signal centered in the IF filter.

At the output of the second mixer is a double-tuned bandpass filter. This filter's pass-band must be wide enough (~1.7 MHz) to pass the C/A code-modulated signal, as at this point the carrier still has the C/A code modulation on it. This filter is used to remove the image frequencies and other mixer products from the output of second mixer.

6.4.3   Correlator and C/A Code Scan/Track Subsystem

The Bi-Phase modulator, 1 kHz bandpass filter, and the Code/Scan block form a system that scans the received signal for correlation with the selected C/A code. By properly controlling the receiver's copy of the C/A code, the C/A code modulation is "stripped off" from the carrier at 10.7 MHz. After code strip off the carrier is now a narrow band signal with the 50 Hz data the only remaining (SV-induced) modulation. The correlated second IF (in frequency) is shown at the output of the 1 kHz wide filter.

The C/A code must be searched or "scanned" looking for the point of correlation, i.e., a sliding correlator (see Appendix A). Once correlation is found, the scan function is turned off and the "track" function is activated. The combination of the BPSK modulator and the 1 kHz bandpass filter form the correlator function. The receiver's C/A code generator drives the BPSK modulator. The Code Scan/Track block adjusts the C/A code phase. Through the use of a feedback control system in the C/A Scan/Track subsystem, the receiver's C/A code copy is kept aligned with the received SV C/A code (Track or Lock). The output is fed to a narrow bandpass filter at 10.7 MHz.

A simple amplitude detection method is used to sense if correlation has occurred. This is just the amplitude of the detected carrier after the 1 kHz bandpass filter. The detected signal level is compared to a threshold. If the threshold is exceeded correlation is declared. This detector controls the Scan/Track settings of the Doppler and C/A code Scan/Track subsystems. During the C/A scan, the receiver must also scan Doppler to put the resulting correlated signal inside our 1 kHz bandpass filter.

Once correlation is detected, the C/A code Track function is activated. The Code tracking function uses an error signal to tell which way to move the C/A code to keep the receiver's code locked to the incoming code. The code error signal used in this design is generated by the Tau-Dither method. Appendix A details some common correlation methods and correlation in general.

6.5   Signal Acquisition

From above discussions, we see that Doppler Scan and Code Scan functions are done simultaneously. This process of searching for the correct Doppler offset and C/A code correlation at the same time is called Signal Acquisition. As we have seen both the Code and Doppler require a search to be done. The exact reasons for the search are different.

In the Doppler case, SV movement (primarily) causes uncertainty in exactly where the SV-transmitted frequency will be. For the C/A code even with zero SV movement (i.e., zero Doppler), we would still have to "slide" our copy of the C/A code against the received signal looking for correlation. This is because we do not know the exact delay of signal and we have an unknown error on our clock (i.e., Tbias). Therefore, we must allow our copy to "slip" past the received to "scan" for correlation.

All CDMA systems must do a signal acquisition process to find the code correlation point. But GPS receivers have the added burden of having to scan Doppler while also scanning for C/A code correlation to achieve signal acquisition. It is the author's opinion that the combination of these two systems having to search at the same time is the most difficult part of GPS receivers. It creates a great deal of complexity on top of the already formidable task of just getting (and maintaining) code correlation. Some designs have massive parallel correlators and banks of Digital Doppler filters (FFT) all dedicated to searching the "Code–Doppler Space" for simultaneous correlation in C/A code and Doppler. By using such methods, the time it takes to achieve signal acquisition time is greatly reduced from the modest methods that will be explored here. Even though the methods that will be used here are slow, understanding them prepares the GPS student for the more complex methods.

6.5.1   C/A Code Clock Doppler

As we know from Chap. 3, the code Clock as well as the carrier has Doppler frequency shift. The Doppler present on the C/A code clock is quite small. Even though it is small, it prevents us from aligning our C/A code "once" and then forgetting about it. The Doppler is constantly changing which results in a code clock that is dynamic as seen by the user receiver (assuming zero user reference frequency error). The Doppler on the received code clock and receiver clock reference errors force the use of a dynamic feedback tracking system to continually monitor the "quality" of C/A code alignment and adjust as needed.

6.5.2 Substantial Time Is Needed to Search for Code and Doppler Lock

When serial search methods are used to acquire the GPS signal, the process can take quite a bit of time. The worst case is when the receiver does a "cold start." In such a condition, the receiver does not know its position or what time/date it is. If the receiver does not have any information it must start from scratch and search through all possible SV's IDs and combinations of Doppler and Code. Depending on "best Guesses" of which SVs to look for and chance it could take many minutes for a serial search receiver to obtain "lock" on four SVs.

6.5.3 Time Can Be Reduced with Prior Knowledge

If the receiver has some knowledge of approximate position, date and time of day, it can significantly reduce the scope and time of the Doppler and SV search it must do. A reduction in C/A code search time is also possible but requires a very accurate clock. Usually, it is the Doppler search and which SVs are in view that benefit the most from prior knowledge.

6.5.4 Estimate of Signal Acquisition Time

Estimating Signal acquisition time is difficult. We will assume that an SV is known to be in view and we know its ID (C/A code). We must allow the code enough time to "slip" past the received signal checking for correlation. A conservative code slip rate for this search is 16 Chips per second. Therefore, it takes about 60 s to search the entire C/A code. During this time, the movement of LO2 should be approximately 1 kHz. With a 0.5 PPM reference, we have to scan 12 kHz of combined Doppler and LO1 offset. Therefore, worst case acquisition time would be 12 min. The average time would be about half of this or 6 min. This time can be reduced by increasing the code slip rate a bit and scanning less Doppler Offset (use a better reference for LO1). But only so much can be done. The serial Correlation/Doppler search methods are just slow. That is why many modern receivers use parallel methods here to reduce search time.

Acquisition time is also effected by SNR. As the SNR increases faster sweep speeds can be used and still achieve lock. As the SNR decreases, a faster scan rates may result in the point of correlation being missed or a momentary attempt at lock then breaking out of lock. Each type of control loop will have a slightly different SNR threshold where this may occur.

6.6   Data Demodulator

The last step of the second IF processing is the data demodulator. Its job is to demodulate the 50 Hz data stream that is left on the carrier after the C/A code has been removed by the correlation process. Many GPS receivers use a Costas Loop demodulator. A quadrature-based demodulator is used in the receiver presented here.

The output of the data demodulator is the 50 Hz data that we will use to retrieve SV ephemeris, clock errors, time information, etc. But there is a problem; the data demodulator does not recover the 50 Hz data clock. The 50 Hz data clock (i.e., the 20 ms) is needed to properly read the data from the SV and to properly set the 1 s Dial. The reader may make the same mistake that the author did when he first bumped into this problem. That is to assume that the C/A code epoch divided by 20 reproduces the data clock. The divide by 20 does indeed produce the correct *rate* of the data clock (or equivalently and the correct *rate* of our 20 ms Dial). But to establish the correct *phase* of the data clock/20 ms Dial requires a bit more work as we shall see.

6.7   SV Replica Clock Block Diagram

In Chap. 2, a model of the receiver replica clock was presented. In Chap. 3, we saw how the SV data could be used to establish the correct phase of 20 ms and 1 s replica clock Dials. In this subsection, we will revisit this topic to add to our understanding receiver's replica clock. In particular, we will be adding detail to the process of setting the 20 ms and 1 s dials of the replica clock. Figure 6.2 shows our updated block diagram of the receiver's replica of the SV clock. This figure also shows the recovered SV clock with all dials properly reset and at the "all zero" moment on the master clock. Since the SV clock is recovered, we can read the dials (captured by Snap_Shot) and determine the path delay knowing the time received (=0 for this case). We do not need to see the second counting dial information in this example and as stated earlier its assumed to be there. The path delay is calculated for the state shown in Fig. 6.2. For this calculation, we have assumed that we have access to the master clock. In a real receiver, this is only the case when Tbias $\approx$ 0.

As discussed above we see that the 0.977 μs and 1 ms Dials do not have reset logic. These two Dials are automatically set to the proper *phase and frequency* by the correlation process. Receiver correlation on an SV signal is a dynamic process.

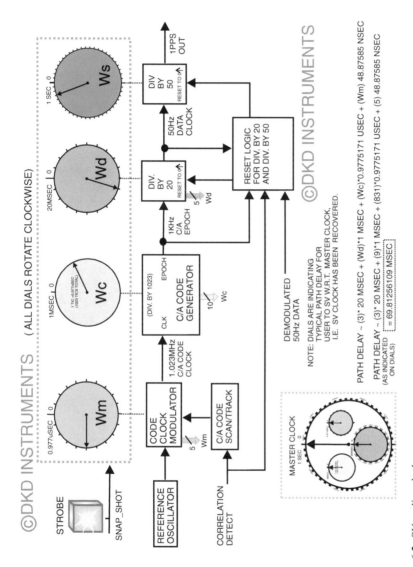

Fig. 6.2 SV replica clock

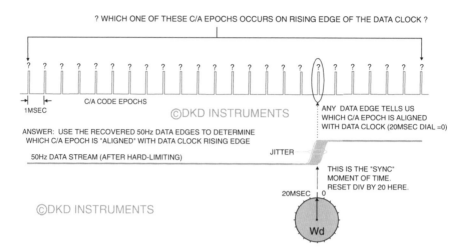

Fig. 6.3 Determining which C/A code epoch is aligned (in time) with the instant the 20 ms dial of replica clock is exactly zero

The result is that these two dials are continually updated by the Code tracking Loop to maintain phase and frequency lock on the received SV signal with its embedded clock signals.

Setting the 20 ms and 1-s Dials must be done by processes outside the code-tracking loop. The most direct way is to use the embedded timing present in the 50 Hz data stream as we saw at the end of Chap. 3. The recovery of the 20 ms dial must be done before the 1-s dial. This is because the recovery of the 1-s dial needs to have the data clock timing before it can find its sync bit in the 50 Hz data stream. It should be clear that the processes of phase recovery of all dials must be done for each SV signal tracked (Fig. 6.3).

6.8   The Data Clock Recovery Problem (The 20 ms Dial of Replica Clock)

The problem of recovering the correct phase of the data clock can be traced to how the divide by 20 is initialized (or reset to zero). Our Recovered SV Clock block diagram of Fig. 6.2 shows a block that is labeled "RESET LOGIC FOR DIVIDE BY 20 and DIVIDE BY 50." The first purpose of this block is to start the divide by 20 counter the "same way" every time we acquire an SV signal. Just turning it on and letting it "free run" will not work. If we do not somehow "sync" this counter to the received SV signal then our regenerated data clock could be off by anywhere from 1 to 19 ms with respect to the received SV data clock. Another way to state the problem is that we need to figure out which 1 kHz epoch pulse occurs exactly at the rising edge of the true data clock. Figure 6.3 shows the C/A code epochs and the data clock. The question is which one of those epochs is the right one to label as occurring on the rising edge of the true data clock?

6.8.1 Recovering the Data Clock Phase (Setting the 20 ms Dial)

We can resolve the data clock phase issue by using the demodulated 50 Hz data and other information available to us in the receiver. The data clock phase recovery process can use the 50 Hz data, Correlation Detect, etc., to issue a "sync" signal which will recover the correct data clock phase. It examines the demodulated data and the divided 1 kHz epoch. If the divider is reset properly, the data edges and the clock edges will be "lined up." This "sync" process only needs to be done once, the counter free runs after that unless the SNR falls so low as to loose lock in the Costas loop. This event would most likely cause the system to break Code Lock so the system would have to reacquire.

The exact implementation of the data clock divider reset (or sync) function can be simple or complex. It could be done completely in software, as the data rate is slow. The best approach would test many data edges before declaring that we have achieved proper data clock phase. A sync method may also estimate receiver SNR before declaring a valid data clock. The reason is as SNR decreases jitter on the demodulated data edges increases.

Many additional pieces of information could be used to develop a robust method for the data clock sync problem. A simple system would use just the 50 Hz data stream and a delayed Correlation Detect signal. Further refinement can be done by including SNR, Data Lock detect, etc.

6.8.2 Noise Effects on Jitter of 50 Hz Data

As SNR degrades we will see an increase in jitter on the demodulated data edges as compared to their true rise and fall times. Figure 6.4 shows the effects. As SNR decreases the observed time of any data edge will depart from true rise/fall time by ever increasing errors. If the jitter on the edges of the data exceeds 1 ms we run the danger of improperly resetting the divide by 20. If this occurs then our 20 ms Dial of our SV replica clock will be incorrect in its phase. A mistake here of one clock tick on the 20 ms dial translates to an error in time of 1 ms. This would of course result in huge errors in computed user to SV distance/range.

Fig. 6.4 Jitter on 50 Hz data
edges

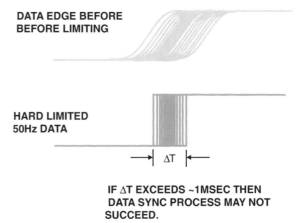

DATA EDGE BEFORE
BEFORE LIMITING

HARD LIMITED
50Hz DATA

ΔT

IF ΔT EXCEEDS ~1MSEC THEN
DATA SYNC PROCESS MAY NOT
SUCCEED.

6.9   Recovery of the Correct Phase of the 1 s Dial

The 1-s Dial of the SV replica clock is not implemented in hardware but would be done in software. It still must be set so as it reads the correct time even though its rate is correct as the dials beneath this level are all locked to the C/A code rate.

The task of properly setting the 1-s dial (or more properly recovering its phase) is quite a bit simpler than the 20 ms dial. The reason is that once the 50 Hz data clock is correctly recovered the timing edges needed are all now well defined and available. This was not the case in the 50 Hz clock case as we had a jitter issue to contend with.

With the correct data clock in hand all that needs to be done to reset the 1 s dial is look for the start of *any* preamble sequence in the data stream as we saw at the end of Chap. 3. The basic issue is really the same as with resetting the 20 ms Dial. We must somehow determine which one of the 50 possible data clock rising edges is "lined up" with the *true* 1-s timing edge that is embedded in the data stream. This is shown in Fig. 6.5. We can use the preamble sequence as it repeats on 6-s intervals. This creates the "virtual" timing mark for our 1-s dial. In other words, the rising edge of the first bit of the preamble sequence coincides with a "1-s" timing edge. By resetting the divide by 50 to zero at this instant, we recover the correct phase of the 1-s dial.

6.10   The Second Counting Dial of the Received SV Clock

At this point all the dials of the received SV clock have been set except for a second counting dial. As GPS counts the seconds in 1 week, resetting Saturday night, we need to finish our SV replica clock by adding a seconds counter. Rather than implement this counter in hardware, this dial and the 1 s dial are implemented in

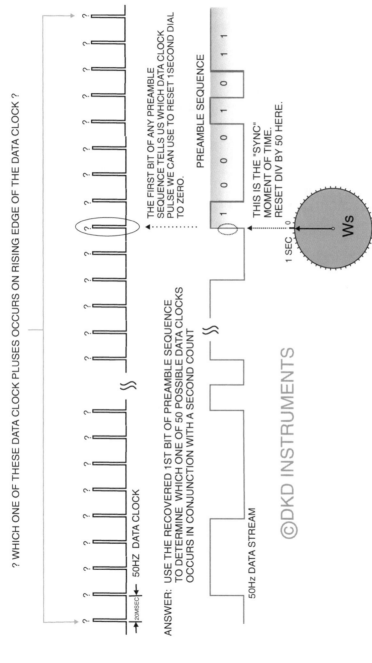

Fig. 6.5 Sorting out which data clock edge/pulse is aligned (in time) with GPS 1 s timing edge in order to set SV replica clock 1 s dial to exactly zero

software. Decoding Zcounts from the data stream easily sets this dial. As noted earlier Zcounts is sent at the beginning of each subframe and refers to the seconds count at BIT1 of the next subframe. The SV clock is now complete and we are ready to take our picture of it using the a strobe called SNAP_SHOT.

6.11 Generating the SNAP_SHOT Signal (Receivers Reference Clock)

A simple way to generate a receiver reference clock would be to just derive a highly accurate 1/10 s pulse train by simple clock division of the 10.23 MHz TCXO. This is the signal called SNAP_SHOT. SNAP_SHOT timing edges occur every 0.1 s and will have an error with respect to true GPS timing edges. This error is called the USER clock bias and can be solved for once enough information is known about user position. This error can be held as a "software error term" or it can be "corrected out" in hardware thereby forcing the SNAP_SHOT timing edges to be nearly aligned with GPS time edges.

The block diagram of Fig. 6.6 shows a receiver reference clock that can be used to generate the SNAP_SHOT signal and modify its *phase* such that the user clock error is minimized. Each time the computer calculates the user clock bias a correction is sent to advance or retard the reference clock, thereby forcing the user clock error toward zero. The exact implementation of the clock modulator will determine the resolution that the user reference clock can be corrected to. The same circuit that is used in the C/A code clock modulator could be used here resulting in a resolution of approximately 50 ns.

6.12 Recording SV Replica Clock Time at SNAP_SHOT Instant

The block diagram of Fig. 6.1 shows the signal SNAP_SHOT as derived from the receiver's reference oscillator at 10.23 MHz. To capture SV time its not necessary to have the 1-s and second counting dials explicitly done as outlined above. Instead, if we know the bit position in the data stream at the snap shot instant we can derive these two dials from this information and Zcounts. For example, as shown in Fig. 6.1 we can calculate an estimate to SV received time as:

$$\text{Est. of SV rec. time @ SNAP.SHOT} = \text{Zcounts(s)} + N \times 20\text{ms} + \text{Wd} \times 1\,\text{ms}$$
$$+ \text{Wc} \times 0.977517\mu s + \text{Wm} \times 48.8759\,\text{ns} - \text{Tsv.err} \tag{6.1}$$

where N is the bit number (counting from bit 1 of any subframe) that the SNAP_SHOT pulse occurs in.

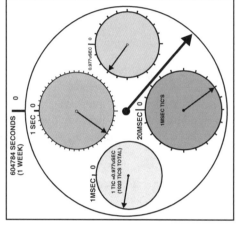

© DKD INSTRUMENTS

TIME ON REC. REF. CLK @ SNAP_SHOT = 226794.6 SECONDS

TIME ON MASTER CLK@ SNAP_SHOP = 226794SEC + (30)*20MSEC + (8) *1MSEC + (792)*0.9775USEC + (17)*48.87NSEC

= 226794.608775024 SECONDS

THEREFORE, REC CLK ERROR IS : Tbias = 8.775024 MSEC

Fig. 6.6 Receiver reference clock relation to GPS clock

6.13   The Data Record Method

In practice, a computer would form a record for every received data bit. At every data bit clock rising edge the data bit would be recorded as a 1 or a 0. Also a check would be made to see if a SNAP_SHOT signal had occurred. If a SNAP_SHOT signal has occurred for this bit then the contents of Wd, Wc, and Wm registers are read from receiver hardware (see Fig. 6.1). Since each data bit is separated in time from the next by 20 ms we can find any given bit's GPS time to within 20 ms by finding Bit 1 of any subframe and knowing Zcounts, as above.

Every time a SNAP_SHOT pulse is received the computer adds 1/10 s to its internal software reference clock. Assuming the computer initially sets its clock to Zcounts plus 60 ms it will at most be off by 20 ms as the delay to any SV is known to be between 60 and 80 ms. Stated another way, we can obtain GPS time to within 20 ms by just getting data from one SV and decoding Zcounts properly.

The data record now has all the information needed to compute the pseudo range. The time received is just the computers count of SNAP_SHOT intervals plus its initial guess of GPS time from Zcount information and the SV received time from the above method.

One complication in this method is the fact that SNAP_SHOT pulse can occur *anywhere* within a data clock period. This follows from the fact that these two signals are asynchronous. The result is the computer may miss assign which bit period SNAP_SHOT occurred in the data record. In order to assure the proper bit time is assigned to SNAP_SHOT instant we can use the data clock divided by 2. By reading this signal every rising data clock edge (and noting its state in the data record) and its state at the instant of SNAP_SHOT, the true location in the data bit record of SNAP_SHOT is assured. The data record allows the computer to resolve the above outlined timing issues.

6.14   Data Processing

If we take four of the data records (as above) from four different SVs we have all the information needed to solve for user position and user clock error/bias. In a single-channel receiver, one SV at a time is tracked and recorded. The slower pace of this method puts more pressure on user clock quality (in our case the 10.23 MHz TCXO time base). A user clock that drifts over the time it takes to do four SV measurements will contribute errors to computed position and user clock error. In short high speed, multi-channel designs have an advantage in that they can complete a measurement in much shorter times. This results in less time for the user clock to drift, which means a less expensive clock can be used in the receiver.

6.15 Absence of AGC

This receiver does not have an AGC control system. In general AGC in CDMA systems is troublesome, as the signal is not apparent until correlation occurs. The GPS signal has very low dynamic range. Typically, a power fluctuation of less than 5 dB is seen for a clear line of sight to SV. The low dynamic range permits building a receiver that can operate well without AGC.

6.16 Summary

This chapter has painted a broad overview of the hardware portion of a four-channel GPS receiver. Some of the important technical problems such as Doppler Scanning, Signal Acquisition, and Replica Clock recovery have also been discussed. The details of making the SV received time measurement were discussed. We are now ready to move to the more detailed discussions of receiver hardware.

Chapter 7
Functional Implementation of a GPS Receiver

In this chapter, we will explore the details of the GPS100SC hardware. Each block of the receiver will now be individually discussed in detail. Details of circuits will be presented for most of the blocks. For the Code, Doppler, and Clock recovery circuits some analysis is done along with detailed operational issues.

7.1 RF Conversion to First IF

This subsection will cover the circuits that can be used to convert the GPS carrier at 1,575.42 MHz to the first IF at 46 MHz.

7.1.1 Antenna and Pre-amp

Figure 7.1 shows a Quadrafilar helix design for a GPS antenna. This antenna has excellent gain pattern, VSWR, and frequency selectivity for use with GPS. It is also circularly polarized. This antenna can be made from a single SMA connector, a small piece of 0.085″ semi-rigid coax, some bits of copper wire, and a small chunk of copper rod. If you build one, make sure you twist it correctly to insure correct polarization.

The gain pattern is slightly less than half hemisphere. The resulting gain in dB over isotropic is approximately +5 dBi on axis. The total gain of this antenna over a 3 dBi linear polarized antenna is approximately 5 dB. This increase in total gain follows from a +3 dB increase from circularly polarized over a linear polarization and +2 dB from gain pattern. This is not insignificant given the low power level of the GPS signal, see Fig. 3.5. For more details on such antennas, the reader should consult ARRL antenna book, 15th edition.

D. Doberstein, *Fundamentals of GPS Receivers: A Hardware Approach*,
DOI 10.1007/978-1-4614-0409-5_7, © Springer Science+Business Media, LLC 2012

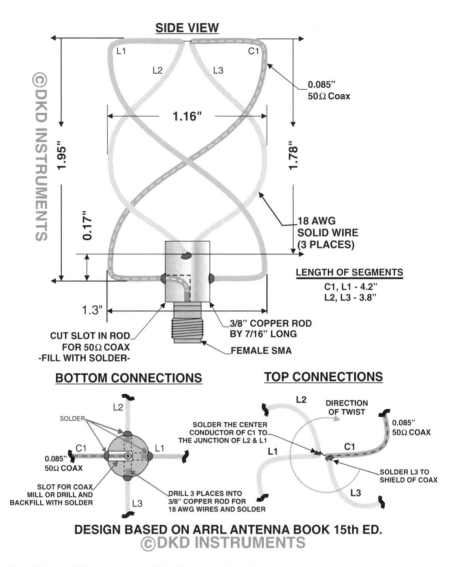

Fig. 7.1 Quadrifilar antenna 1,575 MHz design detail*

No specific details will be given on the pre-amp design. Many low-cost antennas that have integrated pre-amps are now available for GPS. The DC is fed through a bias tee to the pre-amp via the COAX feed cable. Many excellent designs for pre-amps can be found in the HAM radio publications for the 1,296 MHz band. Most of the 1,296 MHz designs can be "tweaked" to give nearly the same (or better) performance @ 1,575.42 MHz as delivered at 1,296 MHz.

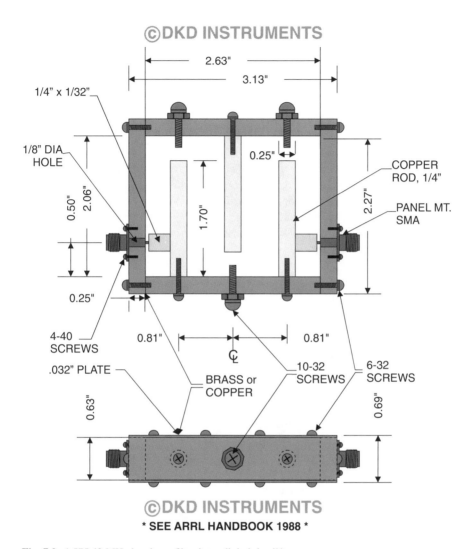

Fig. 7.2 1,575.42 MHz bandpass filter inter-digital detail*

7.1.2 1,575 MHz Bandpass Filter

Figure 7.2 shows an interdigital bandpass filter "cut" for a 1,575 MHz-center frequency. The construction is of brass or copper plate. The tuning screws allow for precise center frequency adjustment. This filter has excellent frequency characteristics. The 3 dbBW is approximately 10 MHz. The Insertion loss is approximately 1.5 dB and the 3 dB bandwidth is approximately 35 MHz.

This filter can be used directly after the pre-amp to provide image rejection at the first IF. It also provides rejection against strong out-of-band signals that could overload the first mixer.

There are of course much smaller filters available that are precut for the GPS frequency. These filters are usually a ceramic resonator type or SAW filters. The filter presented were allows the reader to make this filter themselves.

7.1.3 First Mixer, 46 MHz IF and Filter, IF Power Splitter

Figure 7.3 shows the first mixer and first IF bandpass filter. The mixer is a double-balanced diode mixer, level 7 type. These mixers are available from many manufactures and have decent performance for the task at hand. The output of the mixer is bandpassed with a double-tuned filter tuned to 46 MHz, the first IF frequency. The 46 Mhz filter has a BW of approximately 5 Mhz and an insertion loss of approximately 3 dB. This filter suppresses the image frequency for the subsequent second IF stage.

After the bandpass filter, the signal is split by a signal splitter. The result is that the 46 MHz IF is split two ways enabling a two-channel receiver. The split is of the −3 dB type. The splitters can be made by a few bifilar turns on a ferrite bead or small commercially available Balun transformers can be used.

At the 46 MHz, the signal bandwidth needs are unchanged at approximately 1–3 MHz. The down-conversion process does not effect the C/A code or data modulation, they are intact at this point. What is affected is the Doppler information on the carrier. The Doppler information is corrupted by any errors in the frequency of LO1 are translated onto the carrier now 46 MHz IF. If LO1 were perfect than the 46 MHz IF would have a perfect translation of the Doppler information to this new carrier frequency. In other words, a Doppler offset magnitude of ±5 kHz exists @ 1,575.42 MHz would also exist here @ 46 MHz.

7.2 Second Converter to 10.7 MHz IF

The second converter, second LO, and Correlator are shown in Fig. 7.4. The IF signal at 46 MHz enters this converter as a wide band signal. When it leaves, it is a narrow-band signal assuming correlation is a success.

7.2.1 Mixer and VCXO Removes Doppler Offset

At the second mixer, using a VCXO as the second LO compensates for the Doppler. A crystal oscillator is used as a high-quality, stable signal is needed here. An open-loop LC oscillator or even a ceramic resonator based oscillator would not have enough stability at the frequencies used here. The VCXO is commanded by a DAC that is located in the Doppler Control block. The details of the Doppler Scan/track function are covered below.

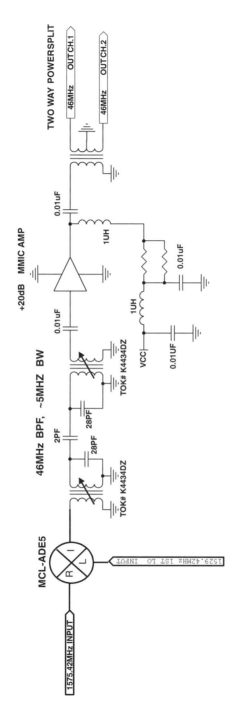

Fig. 7.3 First conversion, first IF bandpass, AMP, and power split

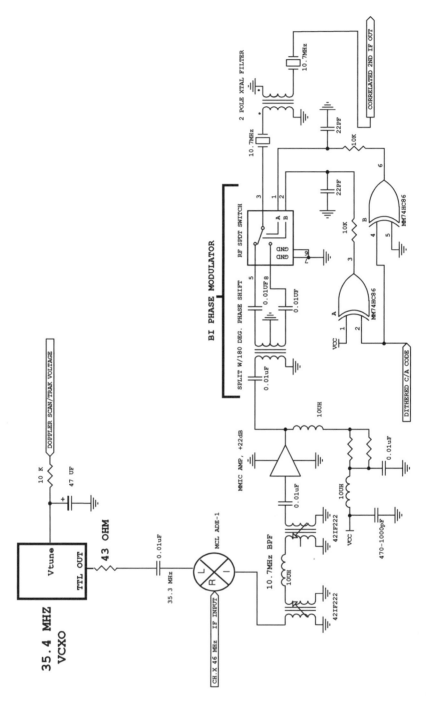

Fig. 7.4 Second conversion, second LO, and corellator

7.2.2   10.7 MHz BPF and Amp

Once the signal leaves the second mixer, it has most of its Doppler stripped off, but it still has C/A Code and Data modulation on it. A bandpass filter follows the second mixer to select the lower sideband from mixer output and to suppress any other unwanted mixer products. Our signal at this point is still a wide band signal at the second mixer output. The 10.7 MHz bandpass filter must be wide enough to pass the C/A code modulated signal without significant distortion. A two-pole L/C filter is used for this filter with a 3 dB bandwidth of about 1 MHz.

An amplifier follows the 10.7 MHz filter providing about 20 dB of gain. Putting the amplifier after the BPF is done so as to minimize its exposure to undesired signals present at the second mixer output. This is an off-the-shelf MMIC amplifier with 50 ohm impedance on the input and output. The amplifier also severs as an isolation element. The following stage is the Correlator. It is good to have isolation from the correlation process, as it is rich in frequency content.

7.2.3   10.7 MHz Correlator with Crystal Filter

This is where the received signal gets the C/A code "striped off." A simple switch based BPSK modulator is used, see Appendix C. The modulator is fed the receivers replica of the C/A code for the particular SV the that is to be tracked. If the replica code is correctly aligned in time with the received C/A Code the BPSK modulator will "undo" the C/A code modulation on the received signal. Once this is correctly done and the C/A Code dynamically tracked, we have a narrow-band signal with only 50 Hz Data Modulation remaining.

The bandwidth after correlation is approximately 100 Hz. But the 10.7 MHz crystal filter following the BPSK modulator has a bandwidth of 1 kHz. As explained above, this is necessary as we have Doppler uncertainty until the Doppler tracker "locks" onto the correlated signal. By having a bandwidth of 1 kHz we can reduce our Doppler search time yet still have high enough SNR's for the C/A code acquisition and track loop to do its job and lock on to the C/A code from the SV.

The two-pole crystal filter following the Correlator has an out-of-band suppression of about 30 dB maximum. This is not very much. But at this point in the processing, we have quite a bit of rejection outside about 5 MHz window about the carrier. This is from "upstream" bandpass filtering. The modest rejection is good enough as it turns out. The crystal filter needs to see about 50 ohms in and out. That explains the 50 ohm resistor to ground (see Fig. 7.5) on the output of the filter. The input to the SA615 circuit is high impedance. With the terminating 50-ohm resistor and a couple of capacitors a reasonable match to SA615 is obtained.

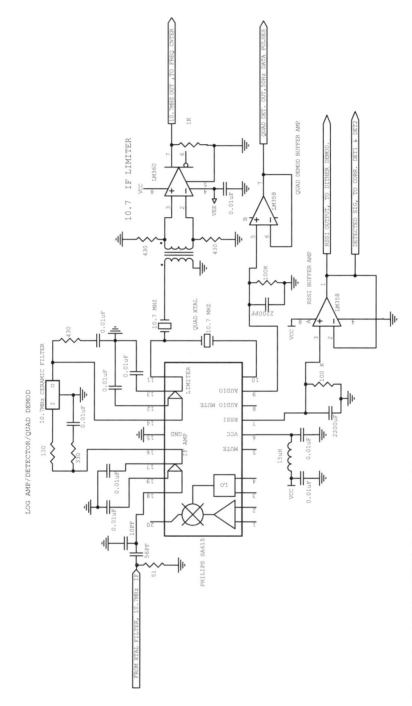

Fig. 7.5 Log AMP, RSSI DET., limiter and quad detector

7.3   10.7 IF Processing Using SA615

Now that our signal is correlated and is a narrow band, we are ready to enter the last phase of the Analog signal processing. Figure 7.5 shows the how the SA615 is used to provide Amplification, Limiting, FM detection, and RSSI level demodulation all at 10.7 MHz.

7.3.1   SA615 IF Processor

The SA615 is a very versatile chip. It was designed primarily for FM demodulation but here we use it for other purposes. We do not use all of the functions available in the SA615. There is an Amplifier/Mixer/LO section that is not used. The IF side is used to do all we need from this chip. It is here that Limiting, Quadrature detection, and RSSI functions are performed. All this is done at the 10.7 MHz IF frequency. There is a ceramic filter used between the IF amplifier stages. This is a wide bandwidth filter associated with FM broadcast. It is typically about 300 kHz wide. It adds some rejection but at this stage the crystal filter has set the IF bandwidth to about 1 kHz.

7.3.2   Correlation Detection and Demod of Dither AM Using RSSI

The RSSI function of the SA615 forms the heart of the correlation detection and tau-dither based C/A code-tracking loop. RSSI is log-scaled amplitude demodulation output. It gives a voltage output that is proportional to the signal level input scaled in dB. The log scaling is really not used here, it is the AM detection we are after. The amplitude of the signal at this point contains two vital pieces of information. First is the correlation detection. If the signal level is high enough, correlation is declared and the C/A code loop goes from "scan" to "track" mode. A simple threshold detector is used for this operation.

The second bit of information RRSI provides us with is the tau-dither induced AM. This signal is induced on the 10.7 carrier to enable C/A code tracking. The RSSI detection provides the first part of the recovery of the C/A code tracking error loop signal processing. Before either detection or tau-dither signal processing is performed, the RSSI output passes through a simple lowpass filter and then to a buffer amplifier.

7.3.3   Quadrature Detection of 50 Hz BPSK Data

The Quadrature detector portion of the SA615 is normally used for FM detection with a LC or ceramic Quadrature element. Here, the quad demodulator is used to do the first stage of the signal processing to demodulate the 50 Hz data still present on the 10.7 MHz carrier at this point.

The 50 Hz modulation is very narrow for a 10.7 MHz carrier. To get more sensitivity at 10.7 MHz, a quartz crystal is used to provide the needed 90-degree phase shift of the carrier for Quadrature demodulation. If a LC tank was used, the resulting signal output level would be too small for subsequent processing still to be done on this signal.

Before that processing can be done, the demodulated signal it is lowpass filtered by a simple RC network and sent to a buffer amplifier. Finally, the Quadrature demodulation method works well with FM signals and gives a immediately useful signal. For BPSK 50 Hz data, we get an impulse at every data edge, not a pulse or data bit. This occurs due to the fact that FM is the time derivative of phase. This means that further processing of the Quadrature demodulated output signal is needed to complete the data demodulation process, see below.

7.3.4   Limited 10.7 IF to Frequency Counter

Even though the SA615 has a limiter output at pin 11 its level is low and the limiting a bit soft. A TTL-level limited signal is needed to drive the frequency counter used for the Doppler TRACK function. A LM360 limiter with a balanced feed does the conversion of the 10.7 MHz to TTL levels. A 10.7 MHz crystal acts as a single pole filter to limit the broad band noise hitting the input of the limiter. The balanced feed on the input stage helps with isolation, improves the sensitivity of the LM360 and helps reduce common mode noise. Care must be used in PCB layout as its easy for the high-level 10.7 MHz to find its way into the SA615 input and desensitize the IF strip.

7.4   Doppler Scan Track Subsystem

Figure 7.6 shows the Doppler Scan and Track subsystem. Most of the system shown resides in a CPLD logic chip. This system is used to scan the range of frequencies where the GPS signal can occur at the 10.7 IF. It does this by sweeping the 35.4 MHz VCXO with a saw-tooth waveform. The sweep function is implemented with 10 Bit PWM DAC being fed by a 10-bit counter that is incremented at the Scan Rate. This produces the Scan ramp and when the counter overflows, it resets to zero starting its ramp all over again.

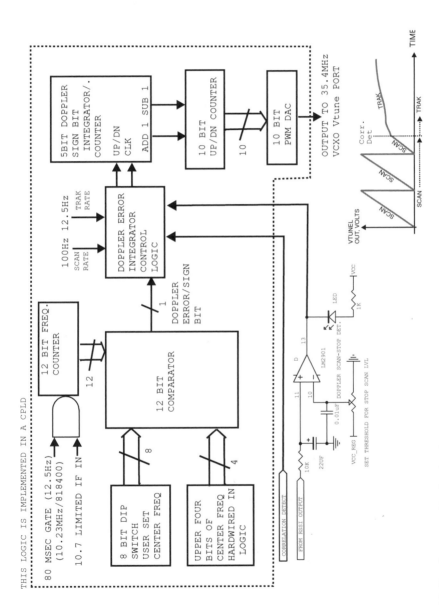

Fig. 7.6 Doppler scan and track subsystem

7.4.1   Frequency Counter Frequency Discriminator

A 12-bit frequency counter measures the frequency of the 10.7 MHz IF to determine if it is above or below the user-entered IF center frequency. The 12-bit comparator now produces the Doppler Sign Bit. Once the Track mode is entered, the control of the VCXO voltage is now closed loop with direction determined by the integrated value of the Sign bit.

The frequency counter is gated by a signal derived from the 10.23 MHz TCXO. This is important. A low drift source is needed for the count to remain unchanged with temperature or age for a constant applied frequency value. Getting the signal to stay inside our 1 kHz wide filter at 10.7 MHz is not easily done. Any drift in the gate width time interval will translate to an error in the frequency count. This in turn will translate into an error in the position of the recovered GPS signal within the 10.7 crystal filter.

Lastly, it should be noted that the data modulation is present on the frequency being counted. This will cause a count variation from cycle to cycle. A digital filter (see below) is used on resulting Doppler Sign Bit before a decision is made on which direction to move the VCXO. This reduces the effect these count variations have on the final commanded voltage to the VCXO.

7.4.2   Center Frequency Control

The center frequency is supplied as a 12-Bit word to one port of the comparator. The upper 4 bits of the center frequency select are hardwired into the CPLD. A DIP-switch enters the lower 8 bits of the 12-bit center frequency select word. This allows for variations in the center frequency of the 10.7 two-pole crystal filter and the Quadrature detection crystal.

7.4.3   Digital Doppler Loop Filter

The sign bit is not passed directly to the 10-Bit UP/DN counter feeding the DAC. Rather it is "integrated" using a 5-Bit UP/DN counter. This ensures that the average of the Doppler Sign bit is acted on, not the moment by moment variation in the sign bit. The Track rate clock samples the Sign bit. If 32 sampled Sign bits occur in the UP direction, an ADD pulse is generated. A subtract pulse after 32 DN sampled sign bits. The Sign bit sampling and integration ensures smooth operation of the Doppler tracker. To implement the Scan function, the Sign bit is held low (or high) thus generating a ramp at the output of the DAC. This is how the SCAN ramp is generated with the control of this determined by STOP-SCAN bit.

7.4.4   Level Detection and SCAN/TRAK

The RSSI function of the SA615 is used to provide the signal level information for Correlation Detect and STOP_SCAN detect. Once correlation signal level is detected, it sets into motion the change from Scan to Track in the C/A code Scan/Track subsystem. The Doppler subsystem does not enter track until the input signal level rises above the threshold set for Correlation Detection, see below. The DOPPLER SCAN_STOP circuits are shown in Fig. 7.6 sets this threshold. This allows the scan rate, which is eight times faster than the track rate, to continue until the signal frequency is closer to the center of the 10.7 MHz crystal filter. If the Track signal level is achieved, the system response rate is dropped by a factor of 8. Once in Track the changes are made to the VCXO are very slow as it is now tracking the Doppler on the GPS signal. These two rates are shown as 100 Hz (Scan) and 12.5 Hz (Track).

7.5   Code Tracker

The C/A code tracking system is the heart of this GPS receiver. With the ability to locate C/A code alignment and continually track received code movement in time the GPS signal is "recovered". Before any other system can "kick in," we must achieve and maintain C/A lock. Figure 7.7 shows a block diagram of the complete subsystem. Other figures will show more details of this subsystem.

7.5.1   Tau-Dither Code Lock

The code-tracking method used in the GPS100SC receiver is the Tau-Dither method. This is a tried and true method that needs only one correlator for both tracking and signal recovery. The basic principles of tau-dither are covered in appendix A. The waveforms depicted in Fig. 7.7 are idealizations of the actual signals present. But they are what you will see if you probe the GPS100SC with the appropriate instrument. The detected correlation pulse can only be observed if the loop is opened up and allowed to freely scan the C/A code.

To observe this pulse, just "hold down" the SCAN/TRAK line and observe RSSI signal on an O'scope. It is best to have some sort of simulator to allow easy access to a GPS signal that one can easily configure in terms of level and C/A code.

Overall, the system is hybrid of digital and analog methods. The C/A code control loop is analog up to the Bi-Phase modulator/correlator and the limiter at the output of the active bandpass. From here, the on the loop is digital. The code

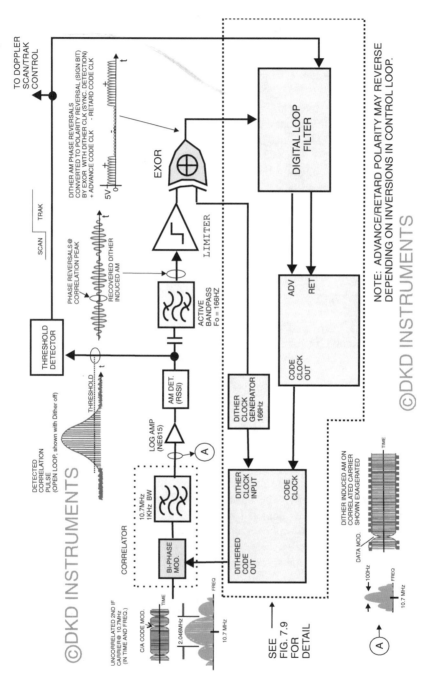

Fig. 7.7 CH. X block diagram of code scan/trak sub-system using TAU-DITHER error generation

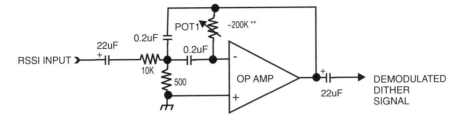

**ADJUST THIS RESISTOR TO BRING ZERO CROSSINGS SYNCRONOUS
WITH DITHER CLOCK SIGNAL, SEE TEXT.

Fig. 7.8 Active bandpass filter circuit

clock modulator and the loop filter were both implemented as digital functions to
ease part count and give a more modern feel to the design.

7.5.2 *EX-OR Detection of Code Error*

The Exclusive OR gate is used to multiply the hard limited demodulated Dither AM
signal with the Dither Clock signal. This is an approximation to a full analog
multiplication process. As a result some information is lost. The information
retained is the SIGN of the C/A Code tracking error. What is lost is the magnitude
of this error. At this point the loop control signal is purely a digital quantity. By
sampling the Code Sign bit at a high enough rate and integrating these samples, the
loop can be closed without significant degradation. In short, the loss of the "magnitude of error" information for the C/A tracking loop is not an issue.

7.5.3 *Active Bandpass Filter Recovers Tau-Dither AM Signal*

The dither induced AM must be recovered to retrieve the C/A code tracking Sign
bit information contained in this signal. The induced AM is a very low level on
the 10.7 MHz carrier. The active bandpass shown in Fig. 7.8 provides gain and
frequency selectivity to extract the dither AM.

This filter is a single pole, active, bandpass filter tuned to recover the 166 Hz AM
signal induced by the dither operation. The RSSI signal is applied to this filter
and the output is a sinewave type signal. If observed on O'scope the signal has a
random envelope structure, which "pulses" the sinewave, see Fig. 7.7 graphic.
At the minima of the output signal, the sinewave is undergoing phase reversals of
180°. It is the phase reversals that contain the sign bit information for the code-
tracking loop.

It is necessary to adjust the center frequency of this filter to match the dither clock frequency. This is done via POT1. Tuning also must be done for another reason. The output of the filter must be phase correct with respect to the Dither Clock signal. The dither clock modulation travels through many circuits, including the active bandpass covered here before the output is EX-OR'ed with dither clock to recover the code sign bit information. This signal path creates delay, which results in a phase offset of the induced dither with respect to the dither clock signal. In short, tuning the filter for maximum output at input of 166 Hz is not enough. The filter must also be tuned such that the zero crossings of the output occur at the zero crossings of the dither clock signal. Since the signal out is at the same frequency as the dither clock, this tuning must optimize signal level output and at the same time the detection of the sign bit by the EX-OR gate. By changing POT1 not only is the center frequency adjusted but the delay through the filter as well.

This tuning is simple and can be done roughly by taking the Dither clock and using it to AM a 10.7 carrier that is injected into the IF (C/A code off, CW mode for receiver). This results in a small error in the tuning due to some delay issues, but it is close. A better method is to use a signal simulator that has the C/A code present. Allow code to lock and then tune the filter by observing the inputs to the EX-OR gate. When there zero crossings are occurring at nearly the same instant of time the tuning is done.

7.5.4 Digital Filtering of Code Error Sign Bit

As we just mentioned, we need to integrate the Code Error Sign bit before we apply it to the control point, the Code Clock Modulator. We can integrate this error bit just as we did in the Doppler case using a UP/DN counter. In this case, an 8-bit counter is used. Figure 7.9 shows the details of digital portion of the C/A code generation and control system. The Sign bit integrator is on the left-hand side of Fig. 7.9. With 8 bits of count we will need the Code Error Sign bit to occur in the UP or DN State for about 256 counts (on average) before a decision is passed along to the Code Clock Modulator. Once this occurs, an ADD or SUB pulse is generated that advances or retards the C/A code by approximately 1/20 of chip.

Two integrating clock rates drive the code loop digital filter. The higher rate, 64 kHz, is for the Scan mode. As in the Doppler loop, we can Scan for C/A code alignment by just forcing the sign bit UP or DN and holding it there. This forces continual ADV (or RET) pulses to be sent to the Code Clock Modulator. The effect is a sweep of the receivers C/A code replica past the incoming or received SV code. This is the C/A code scan mode.

The lower integrating clock rate is for the C/A code Track mode. In very many ways the integrating clock serves as a digital way to adjust the loop bandwidth of the C/A code track loop. Faster clock rates can track a more dynamic received C/A code movement. It is instructive to calculate the maximum rate out of the loop filter/

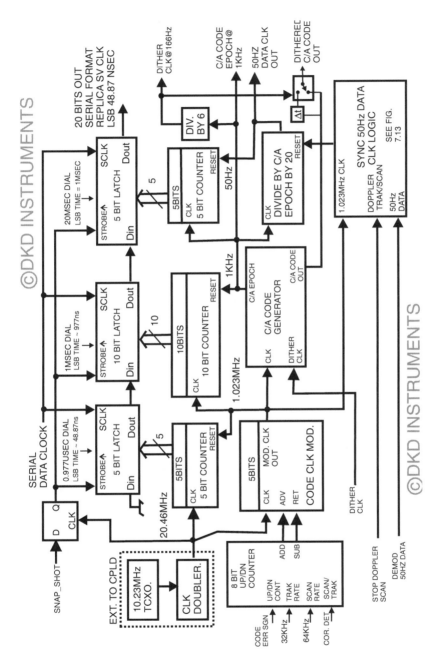

Fig. 7.9 Code clock MOD, code generator, SV replica clock, and code loop filter

clock modulator if we hold the code error sign bit in one state. This is will result in a frequency difference from the nominal code rate of 1.023 MHz.

$$\text{Code clock freq. difference from nominal w/constant sign bit} = [32\,\text{kHz}/(256)]/20 = 6.25\,\text{Hz}$$

$$(7.1)$$

The factor of 256 comes from the sign bit integrator. The factor of 20 comes from the Code Clock modulator. The maximum Doppler rate that the loop can "correct out" of the C/A code loop is about 6 Hz. We know from our Doppler discussions that Doppler as seen by the receiver affects the received C/A code rate. At the maximum, the received code rate is off from 1.023 MHz by about 3 Hz. With a maximum tracking rate correction of 6 Hz, we can easily track the received C/A code.

7.5.5   Code Clock Modulator

The code clock modulator is a digital version of a VCXO. It is an integral part of the C/A code Scan and track system. It allows the receivers replica C/A code to be moved in time with respect to the incoming C/A code received from the SV. The code clock modulator is shown in Figs. 7.7, 7.9 and 7.10. Figure 7.10 shows its internal workings.

In operation, the modulator allows a discrete interval of time to be added or subtracted from the 1.023 MHz code clock. If this interval is continuously added or subtracted at a constant rate, the output frequency is changed. This process is inherently a phase modulation process. If the Phase is changed in a ramp-like manner (steps to be precise), we can approximate the change as a constant offset in frequency. This follows from frequency being time derivative of phase.

The exact amount of time added or subtracted is 1/20 of a C/A code chip. This follows from the driving clock rate of 20.46 MHz, which is 20 times 1.023 MHz. The modulator uses a shift register with 21 stages to effect the delay/divide needed. A single 1 is loaded into the register, which is fed back to the input. If the switch is set to 20 the resulting operation is a divide by 20. This where the switch spends most its time. If an ADV pulse is detected, the switch is set to 19 for the next cycle. After that cycle, the switch is reset to 20. The operation is the same for the Retard pulse except the switch setting is at 21.

If we assume a continuous stream of ADV or RET pulses at Frequency FMOD is applied to the modulator, we can write a general formula for the output frequency of the modulator,

$$\text{Fout} = [20.46\,\text{MHz} \pm \text{FMOD}]/20 \qquad (7.2)$$

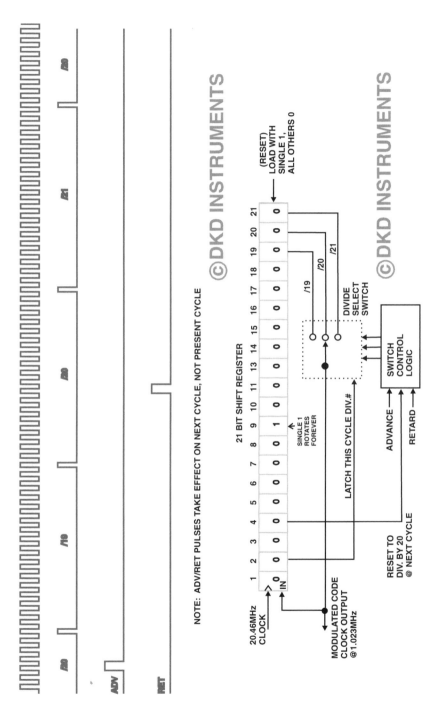

Fig. 7.10 Code clock modulator timing diagram and block diagram

The sign in (7.2) is determined by which type pulse is applied, positive for ADV, negative for RET.

7.5.6 C/A Code Generator, SV Replica Clock, Phase State Counters and Latches

Let's return to Fig. 7.9 and discuss it in more detail. Most of this entire figure is implemented in a single CPLD chip for the GPS100SC receiver. Of particular interest is the C/A code generator portion with its associated counters and latches. It is here that the SV replica clock is implemented and controlled by the receiver. Let's follow the various clock signals as they traverse this subsystem.

The first block it hits is the Code Clock modulator. Here, it is divided by 20 and occasionally 19/21 to enable C/A code clock modulation as we just discussed. We see a counter connected to the Modulator also. The purpose of this counter is to provide phase state information at the modulator. It does this by counting up as it gets 20.46 MHz clock tics. When the Modulator overflows, it outputs a pulse (@ ~1.023 MHz) that clears this counter. This is the hardware implementation of the 0.977 μs dial of our SV clock. A 5-bit latch can capture the state of this counter when it triggered by SNAP_SHOT.

Now let's look at the C/A code generator. It receives the 1.023 MHz clock signal (nominal) from the code clock modulator and essentially divides by 1023 in the process of producing the replica of the C/A code. Rather than capture the state of the shift registers used to produce the C/A code (see Fig. 7.11), a counter is used again. This counter is clocked by the 1.023 Mhz clock from the modulator and cleared by the C/A code epoch. Since the C/A epoch occurs every 1 ms this counter counts to 1023, is reset, and starts over.

This gives a linear readout of the Phase State of the C/A generator. This counter corresponds to the 1 ms dial in our SV replica. As with the code clock state register, the C/A code phase state counter can be captured in a 10-bit latch by the SNAP_SHOT signal.

Lastly, we come to the divide by 20 block. It is here that the C/A epoch is divided by 20. This creates the 50 Hz data clock frequency perfectly, but not its phase. Again a counter is used to easily provide the phase state information about this divider. It is clocked by the 1 kHz signal and cleared by the 50 Hz output clock signal. A 5-bit latch is provided to capture the divide by 20-phase state on the SNAP_SHOT signal. This divider corresponds to the 20 ms dial of our replica SV clock.

As we know form our work in Chap. 6, we must carefully reset the divide by 20 to ensure it is aligned with the true 50 Hz clock (recover its phase) from the received SV signal. Anyone of 20 C/A code epoch pulses could be the correct one that occurs exactly at a data bit edge. The logic of block SYNC 50 Hz DATA CLK provides this function. It uses the demodulated 50 Hz data to determine the proper point to reset the divide by 20, see below.

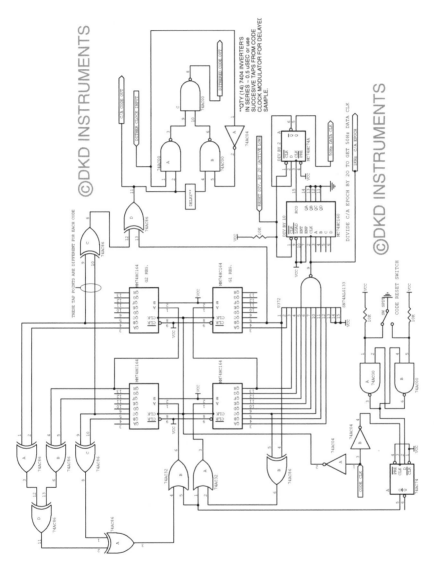

Fig. 7.11 C/A code generator, DITHER code, and 50 Hz data clock generation

The reader may wonder why only the data clock phase must be recovered and not the C/A code clock phase or C/A code generator phase. In short, the tracking of the received C/A code does both of these automatically. In other words, as long as we keep dynamically tracking the received C/A code this process alone recovers the phase of the 0.977 μs and 1 ms dials.

In operation, the phase of the 0.977 dial is never completely accurate. An examination of it would reveal that it jitters as the C/A code is tracked. Nominally, it will jitter about the correct phase by at least 1/20 of a chip. This is of course the resolution of the code clock modulator.

7.5.7 An Example of a C/A Code Generator w/Tau-Dither

An example of a C/A code generator is shown in Fig. 7.11. The C/A code this produces is hardwired for SV 9. A reset button loads the code generator with all 1's, which occurs only once in a complete code cycle of 1,023 bits. Once loaded with all 1's the generator free runs and will continue to produce the chosen C/A code. Combining two 10 bit code generators generates the C/A code. The result is the single unique C/A code for each SV. To change the C/A code to a different SV the tap points for the EX OR gate are changed. It is a trivial task to make this generator support all of the 32 C/A codes with a 5-bit selection logic system. That improvement is not shown here, just the basic generator. For more on C/A code generation see appendix B.

A method is also shown to generate the delayed C/A code needed for the tau-dither code control loop. As shown, a delay element is used. This could be a chain of gates or another delay method such as using tap points from the code clock modulator to sample the C/A code at different times/delays. The delayed code and no delay code are switched at the dither clock rate to form a dithered C/A code output. It is this dithered code that induces the AM on the 10.7 MHz carrier that we pickoff using the RSSI AM demodulation function.

Sensing when the all 1's state occurs generates the C/A epoch. The all 1's state happens every 1023 code bits or every 1 ms. Dividing the C/A code by 20 produces the 50 Hz data clock. The circuit logic to reset the divide by 20 for phase recovery is not shown in Fig. 7.11, see below.

7.6 Signal Acquisition Process

GPS signal acquisition is a complex process. It is comprised of a search or scan phase followed by a transient capture phase, then a lock phase. Also the process is split into Code and Doppler acquisition phases. Many variables come into play. The largest is SNR. Generally speaking, SNR has the most influence over how long it takes to find and lock onto the GPS signal. If the SNR is low the correlation process may not be able to maintain a lock and then lock will break and the scan process will start over again. Larger SNRs allow faster scan rates, therefore faster acquisition times.

Bandwidths also greatly influence the acquisition process. Narrow bandwidths favor the lock process while wide ones favor the acquisition process. An ideal receiver would adjust its bandwidths according to the estimated signal environment the receiver is in. For the GPS100SC receiver, this is not possible. Still by carefully choosing the fixed bandwidths used in the design, it is possible to get good performance from a fixed BW design.

7.6.1 Searching for the Signal, Search Rates, and Alternate Methods

The search for the signal must comprise a Code Search and a Doppler search. Usually these are separate subsystems just as implemented in the GPS 100SC receiver. C/A code correlation must be achieved before Doppler tracking can occur. No Doppler tracking can be done until the C/A code is locked up and the signal level is high enough for Doppler tracking to start.

The GPS100SC has modest search times. Typically, it takes 3–5 min from a reset event to find and lock onto the GPS signal. Code Correlation usually occurs when the signal is near the skirts of the 10.7 MHz crystal filter. The soft roll off of the two-pole Crystal filter helps the capture process. The signal can be captured even when it slightly outside the 3 dB bandwidth of this filter. If a sharp filter were used this could not occur.

The Doppler is scanned at about 14 Hz per second. The total Doppler time to scan ±5 kHz is about 12 min. Code scan rate is set be the 64 kHz-code scan clock feeding the sign bit integrator. The Code is scanned at a rate of about 12.48 Hz/Bit. This works out to about 82 s for one entire C/A code cycle. See appendix A for more information on C/A code scan rates. In a complete Doppler scan there are about nine complete C/A code scans.

All digital designs may use a FFT-based Doppler scan system where all possible Doppler frequency cells can be search from one FFT. This is equivalent to many 1 KHz wide filters, as opposed to the single filter used in the GPS100SC receiver. Each filter would be at a slightly different center frequency. If a correlator were provided for every chip of the C/A code there would be no code scanning needed to find C/A code correlation. A system as described would essentially search all possible code and Doppler possibilities simultaneously. Such a system would have extremely fast acquisition times.

7.6.2 Detecting Code or Doppler Lock and Switching to Track

The switch from scan to track for Code and Doppler is done by simple threshold Detection of the RSSI signal from the SA615. The RSSI is lowpassed filtered for both detector operations. A smaller time constant is used for Correlation detection than for Doppler Detection. This is due to the quicker response time needed for the Code loop. If the RSSI is filtered too much the code scan function may sweep past correlation before the track function is switched on. The time RSSI filter time constant used for the Code detector in the GPS100SC is about 20 ms. With the code scanning at 80 ms per chip the 20 ms-time constant will allow the timely detection of correlation without worry of missing the event altogether.

The Doppler has its own Stop Scan detection so that the Doppler scan process can continue until the signal is nearly in the center of the 10.7 MHz crystal filter. The Doppler Scan stop detection is a much slower process. In this case, the RSSI can be filtered with a much longer time constant, typically 0.5 s.

We need a separate Doppler stop scan detection as code lock will normally occur when the signal is just outside the sweet spot of the crystal filter. By allowing the Doppler scan to continue until the signal rises to a higher level the signal will be pushed into the sweet spot faster as the scan rate is eight times the lock rate for the Doppler loop. For the reasons just covered, the stop Doppler scan level is set at a higher value than the code detection level. Typically, the code detection level is set just above the noise level. The Doppler detection level is set for the lowest level expected for the received signal when it is in the center of the crystal filter. Both Code and Doppler detection levels are adjustable via potentiometers. Finally, the correlation detector circuit is not shown. It is the same circuit as shown in Fig. 7.6 with a different input filter time constant, see above.

7.7 Data Demodulator

The 50 Hz data remains on the carrier after the C/A code is striped off. It can be demodulated in a variety of ways. Many GPS receivers use some form of a Costas Loop demodulator. Costas Loop demodulators can be problematic when implemented in hardware. A simpler approach is used in GPS100SC receiver based on the Quadrature demodulator present in the SA615.

7.7.1 Block Diagram of Data Demodulator, Operation

A block/circuit diagram of the data demodulator is shown in Fig. 7.12. The SA615 Quadrature mixer produces a demodulated FM from the applied carrier. It does this by multiplying the 10.7 MHz carrier by phase shifted version of itself. If the phase

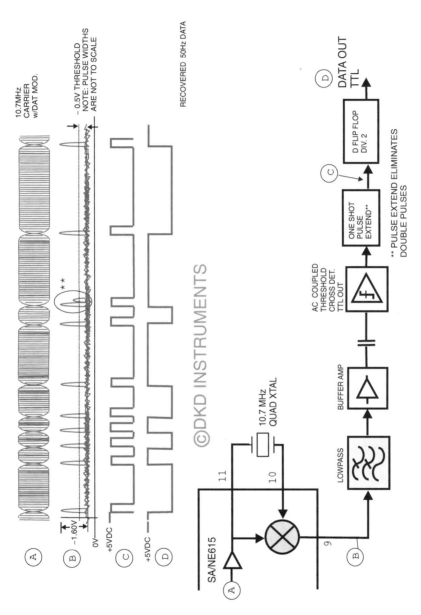

Fig. 7.12 Quadrature method of narrow band BPSK data demodulation using a crystal as 90° phase shift element

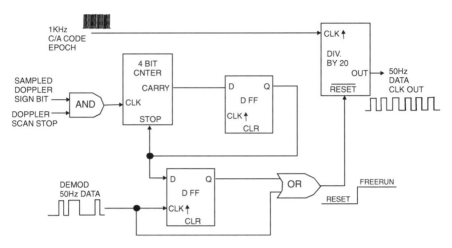

Fig. 7.13 Data clock sync circuit

shift is 90° FM information is recovered after a lowpass operation. The lowpass used is just a resistor capacitor, single pole type.

The response of a FM demodulator to BPSK modulation is an impulse at every bit edge. See waveform B in Fig. 7.12. This is because the FM is the time derivative of phase information. The result is the received 50 Hz data has been pushed through a differentiation process. The pulses must be further processed to undo this differentiation process.

After a simple buffer amplifier, the data pulses are AC coupled to a threshold detector. The AC coupling removes the DC present from SA615 DC bias on the output pin. The threshold detector converts the analog pulses to digital pulses with TTL level. See plot C in Fig. 7.12. A One-Shot pulse extender serves to mask off any pulses after the main pulse.

Finally a divide by two-D flip-flop restores the data by changing state on each data edged. This is another example of simple digital counter (1 bit) exhibiting integration by just counting clock pulses. It is this last operation that "undoes" the differentiation of the FM demodulation process.

7.7.2   50 Hz Data RESET's the Divide by 20 Block

Lastly, the circuit used to reset the divide by 20 of the SV replica clock, see Fig. 7.13. Recall that we must set this divider properly to recover the phase of the 20 ms dial, which counts in 1 ms tic's.

The circuit consists of two parts. First is a circuit that counts sampled Doppler sign bit pulses. This circuit is not active until the Doppler enters the TRACK mode of operation. Once in track and after sixteen SIGNBIT_OUT pulses, the signal is considered "acquired" and the SNR high enough for reliable data demodulation.

Once this occurs the SYNC_DATA clock signal is generated. This signal controls the enable control reset line of divide by 20. When released the divide by 20 starts counting from zero. Until the above conditions are met the release line is held such that every data edge resets the divide by 20. Once the conditions above are met the enable shuts off the data edges and the divide by 20 free-runs from that point on.

Obviously this is a simple system. Many improvements could be made. Regardless, it does a fair job of syncing the 50 Hz clock divider to the timing edges embedded in the 50 Hz data. A better system would average a number of data edges for each reset event. Also, the RSSI would be measured to make a rough estimate of SNR. This would allow a better estimate of when to trust the data from the demodulator and the best time to reset the divide by 20.

7.8 Summary

In this chapter, we took a detailed look at the hardware used to implement the GPS100SC receiver. We started at the antenna and ended at the data demodulator. The hardware of the GPS100SC is very modest by today's standards. The design is not meant to be state of the art but rather a design for learning.

Part III

Chapter 8
GPS Time and Frequency Reception

Most users associate GPS with finding user position. This is of course what many users want from GPS. But a significant and growing application is using GPS to provide time or frequency information. The time information from GPS could be a common clock application of modest accuracy needs or a high precision terrestrial-based reference locked to the GPS clock. The degree of accuracy required for the application will, to a large extent, determine the complexity of the solution needed.

In this chapter, we explore using the GPS L1 signal to receive Time and Frequency information. This is sometimes called Time Transfer. For this discussion, it is assumed that the receiver knows its position and wishes to receive GPS time or frequency only. In addition, DDS-based methods are discussed that allow very precise testing and reference frequency generation.

8.1 GPS Receiver in Time and Frequency, Rate and Phase Errors

When a GPS receiver is configured for Time and Frequency solution it is usually assumed that its position on the earth is fixed (stationary) and known with precision. In this text, we will call such a receiver that reports only rate and phase errors a *Clock Mode Receiver*.

This type or receiver mode enables increased measurement precision of the rate and phase errors of its local reference clock. This increase in rate and phase precision occurs because each received satellite is now an independent measure of receiver clock rate and phase error with respect to GPS master clock. This allows combing or averaging the individual satellite estimates of rate and phase error for increased precision.

GPS receivers solve for phase and rate errors of the local reference clock even during non-stationary or dynamic conditions. But under dynamic conditions, i.e., a moving receiver, the measurements of pseudo range and range rate are used to solve

D. Doberstein, *Fundamentals of GPS Receivers: A Hardware Approach*,
DOI 10.1007/978-1-4614-0409-5_8, © Springer Science+Business Media, LLC 2012

for position as well as receiver clock rate and phase errors. This reduces the precision that can be obtained compared to the when receiver is fixed (and static) in a known location.

In this section, we will use the terms receiver clock, local reference clock, and local clock interchangeably. Modern receivers often have many clocks or oscillators but only one is used as the receiver's reference clock as we shall see.

8.1.1   An Instrumentation Model of GPS Receiver Clock Rate and Phase Measurements

From a user perspective, the Rate and Phase errors reported by the receivers navigation solution (typically available every second in its message stream) can be modeled as measuring the receivers clock phase error with a Time Interval Counter (TIC) and its rate error with a delta frequency counter, with both instruments using a local GPS Atomic 10 MHz clock as a reference, see Fig. 8.1. The GPS 1 PPS time mark is the phase reference input for the Phase error measurement. When the reported phase difference between GPS 1 PPS and the receiver clock 1 PPS is held too small values, typically below ~50 ns, the receiver is said to be reproducing GPS Time.

It is tempting to suppose we could directly hookup to the 1 PPS or 10 MHz signals coming from the GPS atomic clock in Fig. 8.1. But in our instrumentation model we cannot do that. Those signals are not directly accessible. The only information the receiver can provide us is the rate and phase *differences* between the Atomic Clock signals and the Receiver clock signals as displayed on the Time Interval Counter and the Delta Frequency counter.

The example receiver clock is composed of a master oscillator at 16.8 MHz and a clock dial that divides this rate by exactly 16,800,000 to produce a 1 PPS signal. The choice of the master clock frequency of 16.8 MHz reflects a common one used in modern receivers but many others are possible. All of the modeled functions of Fig. 8.1 are contained inside a GPS receiver when it is properly tracking!

The frequency counter operation is a delta between the nominal rate of the master oscillator/clock, 16.8 MHz, and the actual rate as measured against GPS Master Clock rate. In the figure, the reported Rate error, ε, is +101.013 ns/s. The counter scales it to a frequency error delta of 1.6970184 Hz as would be observed at the master oscillator. In other words, if the reported or measured rate error is +101.013 ns/s our 16.8 MHz oscillator/clock is above nominal rate (as referenced to GPS rate) by 1.6970184 Hz.

The time interval counter measures the Phase error of the Atomic clock derived 1 PPS signal versus the receivers 1 PPS output signal. Both modeled instruments use the Atomic reference as their internal clock or reference.

The atomic clock reference is actually the GPS system clock, which is indeed a very accurate and expensive clock. This is the power of GPS receiver in Clock mode; it can report the local clock errors as measured against a very

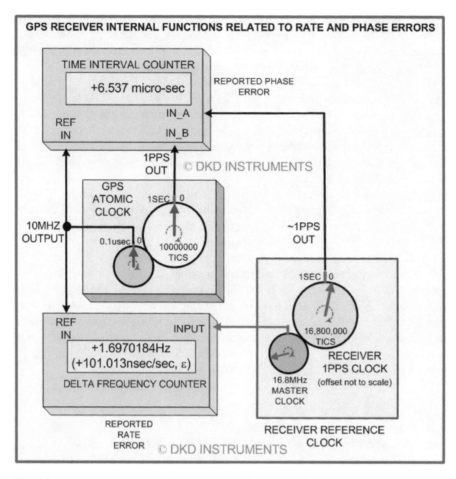

Fig. 8.1 An instrumentation model of receiver clock rate and phase errors with respect to GPS rate and phase

expensive atomic clock in effect transferring that accuracy to the local clock. Often GPS receiver reference clocks are low cost and moderate performance. But if we know the errors of our local clock against a high quality reference clock we can, in principle, correct our local clock to near the precision of the expensive atomic clock.

8.1.2   Reported Rate and Phase Precision and Scale

The precision of the Phase error is less than the reported Rate error for nearly all GPS L1 receivers. This follows directly from the fact that the 1PPS phase errors are tied to Code phase measurements (C/A Code and C/A Chip dials) while the clock rate error is measured using the Carrier Phase dial.

The reported phase error of the receiver's 1 PPS signal is usually given in seconds, but sometimes it is given in micro-seconds. Assuming no cable delays, we should observe this reported phase offset in the time interval counter of Fig. 8.1

The rate error, ε, is typically reported in nano-seconds per second or sec/sec. This is a unit-less quantity and can be used to derive a rate error in Hertz by multiplying it times the nominal rate of the oscillator of interest. For example, to find the rate error of the receiver's reference clock shown in Fig. 8.1, we would multiply the reported rate error, ε, in units of s/s by 16.238 MHz. Another way to think of the rate error is what one would observe on the Time interval counter of Fig. 8.1. At each time interval measurement, one per second, we should observe the time interval counter data change by the reported rate error.

The reported Rate error should have significant digits down to around 0.001 ns/s and the reported Phase error around 5 ns. Note the vast difference in precision as expected as the phase error is based on C/A Code phase measurements while the Rate error is based on Carrier Phase measurements.

A comment here regarding C/A code receivers and P(Y) code receivers is in order. Generally position and measured pseudo ranges are approximately ten times more accurate in P(Y) code receivers than observed in C/A only receivers. This delta in accuracy can be traced to the P code being ten times the rate of C/A code. But for carrier phase measurements both P(Y) and C/A receivers should have nearly the same resolution. The carrier phase resolution in both types of receivers will primarily manifest itself in velocity or speed accuracy. In short, a high quality C/A code receiver should be able to closely match a P(Y) capable receiver in speed and velocity measurement accuracy.

8.1.3 Corrected and Uncorrected Receiver Clocks

Some receivers apply the rate and phase errors to their local clock in an effort to reduce its errors with respect to GPS master clock. More typically, this type of correction is only applied to the phase of the output 1 PPS signal, while the rate error is left uncorrected.

Often the receiver's 1 PPS signal is phase corrected in discrete steps with an uncorrected rate error present. If such a correction was used in Fig. 8.1, we would observe the receiver's 1 PPS phase error progressing at the uncorrected rate error. When the phase error grew to the phase step size, a correction is done and we would observe a step in the receiver's 1 PPS phase.

Internally a receiver may operate perfectly well on an uncorrected receiver clock. In other words, the receiver knows the phase and rate errors on its reference clock and lives with them. To the outside world, such a receiver can provide a phase stepped 1 PPS timing signal that is, on average, corrected to GPS time while *inside* it has a phase error present.

Uncorrected, static, phase errors are typically not an issue for GPS receivers and are easily lived with inside the receiver's navigation solution. Uncorrected

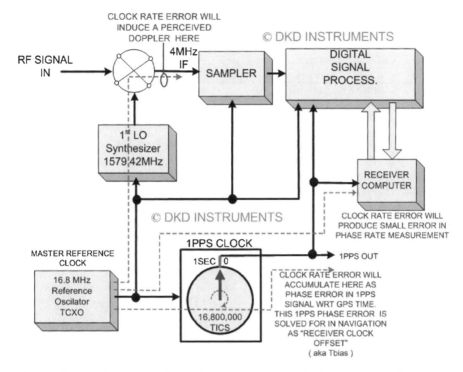

Fig. 8.2 Some of the master oscillator/clock rate error propagation in typical L1 receiver system

reference clock rate errors are a different story. Uncorrected rate errors will accumulate into a dynamic phase error. Static rate errors and slow clock rate drift are usually not a problem for the receiver's navigation solution, but uncorrected *dynamic* rate errors on a GPS reference clock can challenge a receiver's navigation solution. In particular, noise like clock rate errors are particularly difficult to manage. In general, the larger the rate error disturbance, the larger the challenge is. It can very well be that the navigation solution performs better when rate errors of the reference clock are small and slow in movement. Noisy receiver reference clocks can limit Doppler (or Speed) measurement precision.

8.1.4   Typical Receiver Reference Clock System and Rate Error Propagation

Figure 8.2 shows a typical receiver clock. Modern GPS receivers are *coherent*. In particular, all frequencies used in down conversion, all replica clocks rates, measurement signals such as SNAP_SHOT, integration gates, etc. are all tied to a master oscillator or master clock. The rate of these oscillators is usually chosen to be a non-integer multiple of 1.023 MHz for L1 receivers so as to randomize the phase of sampled C/A code clock (i.e., non commensurate).

Table 8.1 Uncorrected rates, periods and measurements due to a receiver clock rate error of +101.013 ns/s, receiver clock is configured in Fig. 8.2

First LO freq	Perceived Doppler due to first LO error	1 PPS period	Measured rate for exact 4.0 MHz IF at sampler input
1,579.420159541952 MHz	159.541952 Hz	0.999999898987 s	3,999,999.59594804 Hz

Figure 8.2 also shows some of the ways in which rate error, ε, on the master oscillator can propagate through the receiver and affect local oscillator frequencies, 1 PPS signal, rate measurements, etc. The rate stability for the master oscillator is closely related to how much it costs and its physical size. Generally speaking, increasing rate stability tracks increasing cost and size of the master oscillator.

The rate errors reported by the navigation solution, as discussed above, are due to the rate errors of the master oscillator. If the master oscillator has zero rate error w.r.t to GPS clock rate, then the reported rate error should be near zero. If you want to find the master oscillator on a typical GPS receiver, try touching some of the components (carefully!) with your fingertip. Usually the heat transfer or other effects will cause the receiver to break lock on all SVs being tracked when you touch the master oscillator.

The total effect of the rate error on all measurements and perceived Dopplers must be accounted for in the navigation algorithm such that not only are true Dopplers calculated (minus Local Oscillator frequency error) but the *measured* Doppler must also be corrected due to small time errors that propagate from the master clock rate error. Specifically Doppler is computed as a change in carrier phase over an interval of time. That interval of time is corrupted by the rate errors on the master oscillator. Table 8.1 shows some of the introduced errors due to a +0.1 PPM rate error on the clock system shown in Fig. 8.2.

8.2   Limits on Estimating Receiver Clock Rate and Phase Errors

Errors in the *measured* L1 carrier rate and the *Predicted* Doppler on the L1 carrier limit the precision that can be achieved in the *reported* rate error of the Receiver Clock.

Carrier Rate measurement errors can occur from many sources. High on list is finite signal to noise ratio (thermal noise/carrier Phase Dial noise), first LO phase noise (Jitter), and IF quantization or sampling noise. Errors in assumed user position, Satellite Vehicle position errors, and atmospheric effects can cause Doppler prediction errors.

Given the above disturbances to Receiver Clock rate error measurement process, one stands out as a primary source, User Position Error.

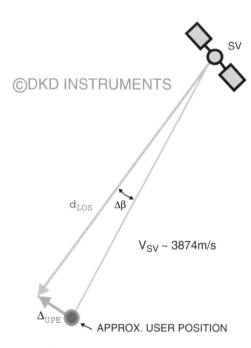

Fig. 8.3 Translating user position error into computed doppler error

8.2.1 Estimating Predicted Doppler Error Due to User Position Uncertainty

The predicted Doppler error limits due to user position errors can be estimated by using the vector method of SV velocity resulting in observed Doppler calculation. This is just the component of the tangential SV velocity vector that is directed at the user receiver, see Fig. 3.9. This figure shows the details of the geometry and some of the equations associated with calculating Doppler using the vector method.

From Fig. 3.9, we see that the velocity component of the SV in the direction of the user is given by;

$$V_\mathrm{d} = V_\mathrm{sv}\sin\beta \tag{8.1}$$

Where V_d is the Doppler velocity as seen by the user receiver and V_sv is the tangential velocity of the Satellite vehicle and β is the angle in radians between LOS vector and a vector from SV to center of the earth.

We will assume that β is very small and now call it $\Delta\beta$, see Fig. 8.3. The angle $\Delta\beta$ is the angle between the true user position and estimated user position. Since $\Delta\beta$ is very small we can use the small angle approximation to the sine function. We also realize that if $\Delta\beta$ is small so is the component of V_d we seek. We now call small V_d, ΔV_d. The relationship expressed in (8.1) now becomes;

$$\Delta V_d \approx V_{sv} \times \Delta\beta \quad (\Delta\beta \text{ must be in radians}) \tag{8.2}$$

If we assume the estimated user position is perpendicular to the line of sight vector from the SV to the user (worst case) we can estimate $\Delta\beta$ as:

$$\Delta\beta \approx \Delta_{UPE}/d_{los} \tag{8.3}$$

Where Δ_{UPE} is the user position error vector and d_{los} is the line of sight distance from SV to user.

Assuming a user position error magnitude of 30 m and the magnitude of d_{los} distance to be 25,000 km (i.e., SV at Horizon, the worst case);

$$\Delta\beta = 30\,\text{m}/25,000\,\text{km} \tag{8.4}$$

$$\Delta\beta \approx 1.2 \times 10^{-6}\text{radians} \tag{8.5}$$

Computing $\Delta V_d \approx 3,874$ m/s $\times [1.2 \times 10^{-6}$ radians], which is about 4.6×10^{-3} m/s

We can convert to Doppler using:

$$\Delta f = [f_{L1} \times \Delta V_d]/C(\text{Speed of light})$$
$$\Delta f = [1,575.42 \times 10^6 \text{Hz} \times 4.6 \times 10^{-3} \text{m/s}]/[3 \times 10^8 \text{m/s}]$$
$$\Delta f = 0.025\,\text{Hz} \tag{8.6}$$

The value of Δf just calculated is one of many accuracy bounds on the predicted L1 Carrier Doppler due to the given errors in user position. The receiver's estimate of the clock rate error will confront this same bound. Assuming we can indeed predict the Doppler on L1 carrier to this resolution we have achieved a receiver clock rate error precision to approximately ± 0.0000158 ppm or $\varepsilon = 1.58 \times 10^{-11}$. As the user position error shrinks, our accuracy will further improve as long as our assumption holds on user position error being the dominant Doppler error source.

8.2.2 Detectable L1 Carrier Phase Rate Limits and Clock Rate Error Precision

Receiver clock rate error estimates are based on detectable phase movements of the receiver's carrier phase dial as observed against a reference dial tied to GPS rate and Phase. For a GPS receiver, the observed Phase movement of the carrier phase dial can be caused by SV movement, receiver movement, Receiver clock rate errors, and by various internal and external noise sources. At some level, the noise sources will obscure the observed phase changes due to SV movement, receiver movement, and Receiver Clock Rate Error. We can estimate this fundamental carrier phase measurement limit with a simple rule of thumb, if we assume the minimum detectable movement of the Carrier Phase Dial is ~1/50 cycle in 1 s of observation.

One L1 carrier cycle at 1,575.42 MHz is equivalent to 0.63475 ns of time. Dividing by 50 gives ~1.26×10^{-12} s. If this phase change was observed in 1 s, the resultant limit in our carrier phase rate measurement is ~1.26×10^{-12} s/s. Our limit in carrier phase rate measurement is also another limit in Receiver Clock Rate Error measurement. This is about an order of magnitude less than the estimate we calculated above for the bound of Receiver Clock rate errors due to a 30-m error in user position. If we assume our receivers clock is perfect, we still would face this same limit in the receiver's navigation solution for Doppler and hence reported receiver speed or velocity. We can convert our Carrier Phase Rate observation limit into an equivalent velocity or speed measurement limit by just multiplying by speed of light;

$$\Delta V_d = \text{Observable Carrier Phase Rate Limit (s/s)} \times C$$
$$\Delta V_d = 1.26 \times 10^{-12} \times 3\text{E}{+}8 \Longrightarrow \text{~0.0004 m/s}$$

8.2.3   Receiver Reference Clock Quality and Rate Error Limits

Two estimates of reported receiver clock rate error precision have been discussed, one based on position errors and one based on carrier phase rate measurement limits. The receiver has another rate error measurement limit and that is the stability of its own reference clock. The stability we speak of here is the rate wandering or rate noise on the receiver's reference clock. This type of clock rate noise is typically expressed as Allan variance or Allan Deviation.

Table 8.2 shows some typical Allan Deviation for some typical 10 MHz reference clocks. Most commercial, low-cost receivers use a TCXO type reference clock. As shown in Table 8.2, clocks of this quality have a frequency uncertainty (or rate wobble) of 1×10^{-9} if averaged over 1 s. If we use this type of clock, it will most likely dominate the receiver's ability to measure receiver clock rate error over this same averaging interval. If we use even a low grade OCXO reference clock, we would see nearly ~3 orders of magnitude improvement in clock stability. In summary for a receiver to hit its inherent rate error limits, as discussed above we need a decent reference clock.

Table 8.2 Typical 1 s Allan variances for different frequency references

	TCXO	Low grade OCXO	Mid grade OCXO	High grade OCXO	Typical Rubidium	High grade Rubidium
Allan deviation, Δf_d @ 1 s	1×10^{-9}	3×10^{-12}	1×10^{-12}	6×10^{-13}	2×10^{-11}	1×10^{-11}
Δf_{L1} ($\Delta f_d \times 1{,}575.42$ MHz)	1.575 Hz	0.0047 Hz	0.0015 Hz	0.00094 Hz	0.0315 Hz	0.0157 Hz
ΔSpeed in m/s (Multiply Allan deviation by C)	0.3 m/s (~0.66 miles/h)	0.001 m/s			0.006 m/s (~0.013 miles/h)	

Notes. GPS receivers like steady, non changing rate, and phase errors on the reference clock. Receiver accuracy in reported clock phase error, velocity, and position is typically enhanced by receiver reference clocks that have high stability in their rate

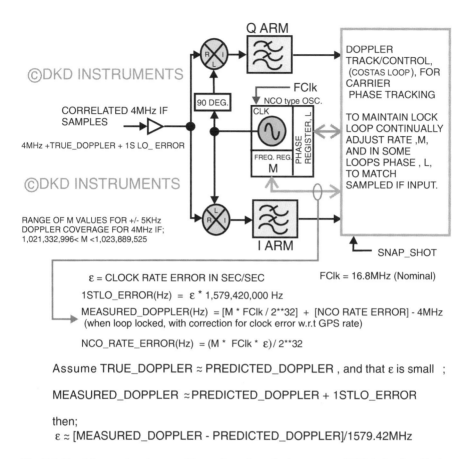

Fig. 8.4 Precision carrier phase tracking and receiver clock rate error, NCO is Replica Carrier Phase e Dial@ IF rate

8.2.4 Measuring Carrier Rate, Doppler and Receiver Clock Rate Error

The GPS100SC single channel receiver presented in Chaps. 5 and 6 is not capable of making precision carrier rate measurements. Very high precision rate measurements can be done using all quasi-digital Costas Loop[1] type carrier trackers. Figure 8.4 shows an example of such a system for use with our receiver of Fig. 8.2. An all digital version would replace the lowpass filters in the I and Q arms digital accumulators and the mixers with multipliers; see Chap. 9. These loops

---

[1] Many receivers avoid the use of a true Costas Loop by avoiding integrating over a data bit edge. But the 180° ambiguity remains due to the 50 Hz data modulation.

often run at update rates near 20 ms and can produce many estimates of Receiver Clock Rate error per second.

Figure 8.4 is a closed loop system that when locked produces a carrier, via the NCO, which is very nearly rate and phase equal to the sampled IF that is applied to its input. *The NCO, of Fig. 8.4, is equivalent to the Carrier Phase dial (down-converted to 4 MHz) of the receiver's replica clock for a single SV being tracked.* Thus the frequency of the NCO (assumed here to be 32 bits) contains a very accurate estimate of True Doppler plus IF Nominal Rate (4 MHz) plus Receiver Clock Rate Error terms. To obtain the True Doppler + Rate Error Terms portion, we must subtract off the nominal IF frequency of 4 MHz.

8.2.5   Estimating Receiver Clock Rate Error

To estimate receiver clock rate error, ε, requires a difference between observed Doppler and Measured or received Doppler for the SV being tracked. Note that for each SV tracked, an independent estimate of Receiver Clock Rate error can be computed.

We can predict the observed Doppler as we know our position (Clock Mode) and we know the SV position and SV orbital rotation rate; see Fig. 3.9. We can obtain a measurement of received True Doppler + Receiver Rate Errors from the carrier tracking NCO as noted above.

The scaled difference between predicted (True) Doppler and Measured Doppler is, to a first order, a measure of Receiver Clock Rate error. As noted above, our measured carrier rate has a small inaccuracy introduced by the Clock Rate error or NCO Rate Error as in Fig. 8.4. With the value of M known and Predicted Doppler, the value of ε can be solved for.

This is a good spot to define the use of term "nominal" with respect to the rates in our carrier tracking system. For this discussion, it means a perfect frequency as would be measured against GPS rate. For example, if we say "Nominal 16.8 MHz", it is meant that that is the exact frequency that would be measured using the GPS Clock Rate as the reference rate. All the specific rates shown in Figs. 8.2 and 8.4 such as 16.8 MHz, 1,579.42 MHz, etc., are nominal rates w.r.t. GPS rate. The actual rates will be in error by the factor ε, as noted and used above.

8.2.6   C/A Code Phase Measurements Limit Time precision in L1 Time Transfer (Clock Mode)

In order to properly position its 1 PPS signal phase near to GPS 1 PPS phase, the receiver must *predict and measure* the delay from the satellite being tracked to the antenna. The difference between these two estimates is an estimate of receiver

Fig. 8.5 C/A code phase
measurement jitter

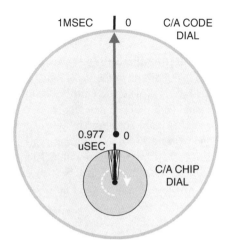

clock 1 PPS phase error with respect to GPS 1 PPS. The final residual phase error of the receiver's 1 PPS signal with respect to GPS 1 PPS signal is limited by errors in the *prediction* of the delay and errors in the *measurement* of the delay.

The primary errors in the prediction of the delay are user position errors, SV position errors, and unknown atmospheric delay.

Code Phase error processes dominate the errors in the measured delay in a L1 receiver. Such a receiver is not able to use the finer resolution Carrier Phase dials due to ambiguity issues. The C/A code-tracking loop has thermal noise in it. This limits the precision that the C/A Chip dial phase can be reproduced from the received SV signals.[2]

The result is jitter on the C/A chip dial as shown in Fig. 8.5. If we assume a limit of 1/100 of a C/A code chip, we get a jitter on the order of 10 ns or ~10 ft using approximation 1 ft/1 ns scaling. The C/A CHIP DIAL phase jitter will ultimately limit the receiver's precision in delay measurement to approximately 10 ns

This limit in delay measurement will be reflected in the final accuracy of the receiver clock phase error as seen on the output 1 PPS-timing signal. For L1 receivers, expect receiver-generated *true* 1 PPS phase uncertainty of 10–20 ns at best. The reader should note that *reported* 1 PPS Phase error is not the same as the phase error *measured* against true GPS time (or a high quality local atomic clock). Specifically, the reported phase error may be quite small but if examined carefully against a local atomic clock one would see phase errors larger than reported. These errors are the physical expression of the errors in the receiver's measurement and prediction of true range delay to the SV.

---

[2] For an excellent analysis of code tracking jitter see Digital Communications and Spread Spectrum Systems, Ziemer and Peterson.

8.3   Initial Estimate of GPS Time

For a user on the earth's surface, the time as seen "looking up" to a GPS SV is always 60–80 ms in the past. This follows directly from the known delay from the SV to users at near the earth's surface. So even with no other knowledge, a single channel GPS receiver can receive the GPS signal and determine GPS time to within 20 ms of actual GPS time by just decoding the received 50 Hz data stream. This is done reading ZCOUNT, setting replica SV clock dials, etc. Once this is done, the User's clock can be set to within 20 ms of true GPS time. To get closer than 20 ms of true GPS time requires precise estimate of the delay between SV and User. This situation is illustrated in Fig. 8.6.

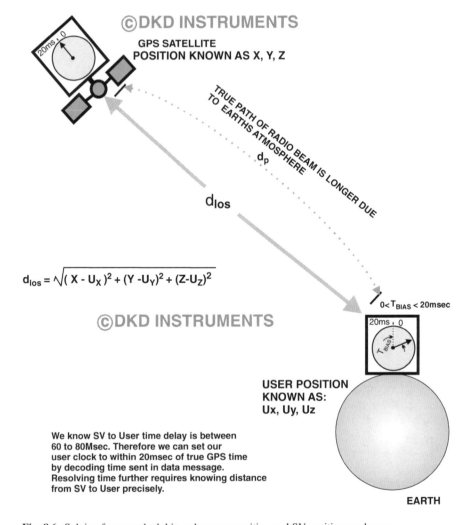

Fig. 8.6 Solving for user clock bias when user position and SV position are known

8.3.1   SV to USER Signal Delay

The two dominant sources of delay between the user receiver and the SV are the LOS path delay and atmospheric delay, see Fig. 8.6. We will ignore relativistic delay effects. The LOS delay is easy to compute, as it is just the LOS distance divided by the speed of light.

Refraction and other processes cause the atmospheric delay. Atmosphere delay is difficult to estimate in the L1 receiver and much effort has been expended in this area. Complex models and measurements must be used to estimate the added delay. For our purposes, here we will not explore these models as many excellent references exist for this. Rather, we will introduce a term, T_{atmo} that we will use to represent this added delay.

8.3.2   Estimating Path Delay

With our known position information, we can calculate the LOS distance directly as:

$$d_{los} = \text{SQRT}[(X - U_x)^2 + (Y - U_y)^2 + (Z - U_z)^2] \qquad (8.7)$$

Where X, Y, Z is SV position in ECEF coordinates and U_x, U_y, U_z are user receiver position in ECEF coordinates. We can compute the LOS path delay by using speed of light;

$$\Delta t_{los} = d_{los}/C \qquad (8.8)$$

C is speed of light

With our estimates, LOS path delay, and delay from atmospheric affects, we can estimate true GPS time by adding this time to the received time from the receiver's SV replica clock minus that particular SV clock error term.

$$\text{Estimate of GPS time} = \text{SV\_CLOCK\_TIME} - T_{err\_sv} + \Delta t_{los} + T_{atmo} \qquad (8.9)$$

Where T_{err_sv} is SV clock error

For the GPS100SC receiver, our SV replica clock would never be better than ± 48 ns due to the code clock modulator resolution. This of course sets a limit also on our ability to accurately reproduce GPS time. Figure 8.7 shows a flow chart of the steps needed to compute and correct receiver clock 1 PPS phase error.

8.4   Verifying the Veracity of Reported Receiver Clock Rate and Phase Errors

The receiver-reported clock rate and phase errors might have errors or misbehaviors under certain scenarios. One way to verify the reported Rate and Phase errors either in Clock mode or Position update Mode is to monitor the Receiver's clock with

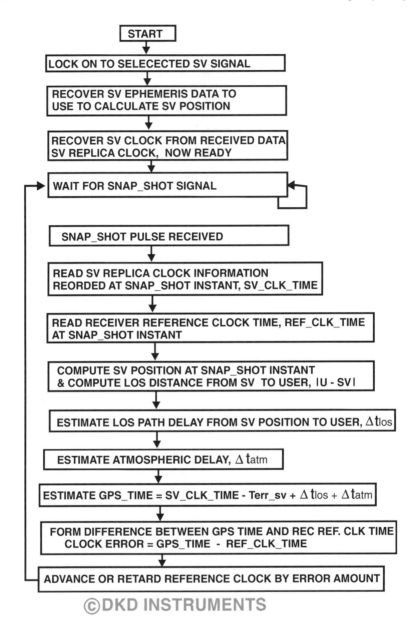

Fig. 8.7 Flow chart for estimating and correcting receiver time to GPS time using one SV

external instrumentation. The idea is to measure receiver clock rate and phase errors with the external instruments and then compare that collected data to the data reported by the receiver at the same time instants. Such a check, properly done, can reveal many interesting phenomenon associated with GPS receiver clock performance and indeed overall receiver system performance.

The precision of the 1 PPS signal (hopefully) and reported rate and phase errors requires quality instrumentation and careful setup. Figure 8.8 shows our receiver model of Fig. 8.1 with time interval counter and frequency counter attached to the receiver clock.

The local atomic clock of Fig. 8.8 should be Rubidium type or better. A Cesium clock is optimum but these are very expensive. A well-behaved Rubidium will have adequate performance and is rate stable over days of operation once it is warmed up.

As shown, the 1 PPS output from the local Atomic Clock is offset from GPS time by approximately 1 ms. This offset insures that the measured phase difference between receiver 1 PPS and Local Atomic Clock 1 PPS should not go through zero and reverse sign. Some TICs can deal with this sign change while others may change scale or worse report an error during the zero crossing data points.

Assuming the Receiver is producing a corrected 1 PPS its offset to true GPS, 1 PPS is usually not of interest but rather its variance around a mean value and excursions from that value is the primary data of interest. As a note to the reader getting a 1-PPS signal in the lab that is coincident with GPS, 1 PPS is fraught with unaccounted delays, which produce phase offsets! Here are some other tips for getting the best data:

- Carefully tap off the receiver's master clock. A buffered 50 Ω output is preferable but a scope probe is also acceptable.
- The counters should have a selectable input termination. If possible, use 50 Ω setting and 50 Ω cables. If a scope probe is used, use the high impedance setting. Keep in mind rise times will change readings. Select thresholds carefully!
- Keep All Cables as Short as Possible
- Offset the 1 PPS from the local Atomic Clock so as to avoid a zero crossing phase error as noted above. The trigger edge polarity may be enough offset.
- Set up the TIC to read the smallest time possible. If we reversed the TIC connections shown in Fig. 8.8, the interval measured is near a second. The longer the TIC counts, the more time errors in the TIC have to accumulate.
- The 10 MHz must be connected to both instruments, especially the frequency counter. If you use the freq counter with its internal reference, you may end up measuring the rate error of the counter instead of the receiver's clock rate error!
- Be careful splitting a single 10 MHz, 50 Ω output from the Atomic Clock. Better to use the 10 MHz output from the instruments in daisy chain fashion.
- Make the Frequency counter the first instrument in the daisy 10 MHz chain. It is the most sensitive to issues that could occur on this reference signal.

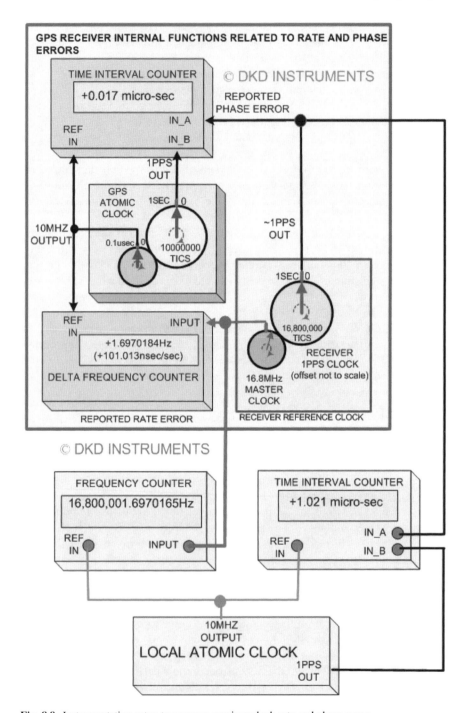

Fig. 8.8 Instrumentation setup to measure receiver clock rate and phase errors

8.5 Using a DDS Based Receiver Clock to Introduce Precise Rate and Phase Errors

Most commercial L1 receivers have limited rate control (if any) of there receiver clock. In particular the Master Clock or oscillator is often a fixed frequency TCXO or for some systems a limited analog rate adjustment is provided. A very useful method for investigating reported rate errors and possibly enhancing 1 PPS output performance is to provide replacement for the receiver's TCXO with a clock based on a local Atomic clock. Standard signal generators do not typically have the precision and resolution needed for such a clock.

An atomic clock-based reference using a 48-bit DDS is shown in Fig. 8.9 and can fulfill our desired needs of precision, resolution, and accuracy. It is shown connected to our receiver of Fig. 8.1 providing a precise, low rate error 16.8 MHz clock signal with extremely fine rate and phase control. The DDS-based master clock shown can be setup for other output frequencies by changing the values programmed into the DDS rate word and the output bandpass filter.

A discrete ×8 multiplier is used to achieve the 80-MHz clock needed by the DDS. Some DDS devices have internal clock multipliers. Often these are PLL-type multipliers with an on chip VCO that is locked to the applied reference frequency. It is the author's experience that often these types of clock multipliers are not suitable for use as DDS clock when the final goal is to drive a GPS receiver master clock. The reason is that master clock (16.8 MHz in this example) is effectively multiplied up to the first LO by a receiver-based PLL – Synthesizer subsystem. This multiplication will increase the phase noise as seen at the first LO. In short, the use of internal multipliers may compromise DDS output phase noise and result in degraded first LO phase noise.

The rate resolution of the 48-Bit DDS output is given by;

$$\text{DDS Rate Resolution} = 80 \text{ MHz}/2^{**}48, \text{ or } 2.842170943 \times 10^{-7} \text{ Hz}$$

This rate resolution would correspond to rate error referenced to 16.8 MHz of;

$$2.842170943 \times 10^{-7}\text{Hz} = \varepsilon \times 16.8 \text{ MHz; or } \varepsilon = 1.691768 \times 10^{-14} \text{ s/s}$$

The rate error resolution limit calculated above is about two orders of magnitude smaller than what can typically be expected from GPS-reported rate error precision.

If we use a rubidium oscillator for the Atomic Reference, it should be possible to set the reported rate error to near zero for many hours and observe the receiver's performance in terms of 1 PPS phase error and reported Rate errors. In addition, a precise step of rate (or phase) can be commanded to the DDS and the reported rate inspected to verify accuracy. Large, precise rate errors can also be introduced to the receiver to verify performance and accuracy. Lastly, the phase control can be used to move the 1PPS signal with typically much finer resolution than many commercial receivers provide in their phase step type corrections.

Combining this system with the independent measurement system shown in Fig. 8.8 enables very sophisticated investigation of Receiver Clock error performance.

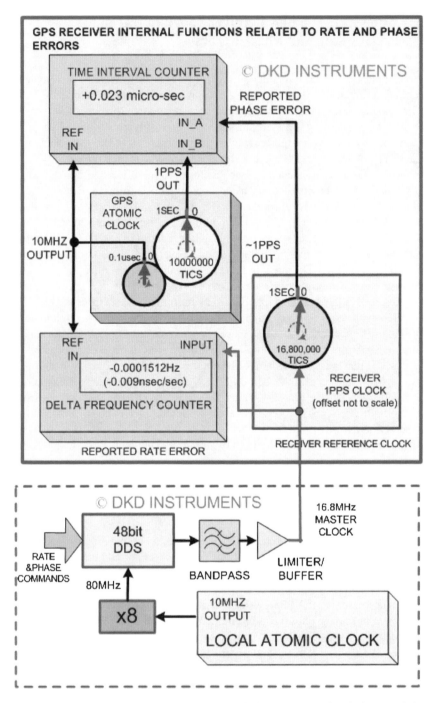

Fig. 8.9 DDS Signal generation of 16.8-Mhz master clock with ultra precise clock rate and phase control

8.6   GPS Disciplined Oscillators

A common subsystem based on GPS receivers operating in Clock Mode is what is called a *GPS Disciplined Oscillator*, or GPSDO. A typical block diagram of such a system is shown in Fig. 8.10 connected to our receiver model of Fig. 8.1.

The primary intent of GPSDO systems is to correct the rate and phase of a high-grade clock (the DO) to GPS Clock rate and phase. The clock is typically at least as good as a single oven quartz crystal type all the way up to a Rubidium type clock. The grade of clock selected is determined by cost and performance issues.

Two modes of operation are typically supported. If the GPS receiver is tracking, the DO is constantly adjusted in rate and phase so as to minimize its rate and phase errors w.r.t the GPS clock. If the GPS receiver is unable to track, the GPSDO system enters what is called *hold over* mode. In hold over mode, the rate and phase of the DO will drift w.r.t GPS Clock. Some GPSDOs may attempt small rate corrections based on aging, temperature or perhaps known rate error performance of the DO in an effort to minimize drift. Desired Hold Over performance will largely determine the grade and cost of the DO used. A common cited specification for hold over performance is total accumulated phase error of the GPSDO 1PPS output (wrt GPS 1PPS) over some interval of time.

The architecture shown in Fig. 8.10 represents a common one where the DO is divided down to a 1 PPS signal and phase compared with the 1PPS from the GPS receiver. The phase error is used as an input to a PLL type control loop which then controls the rate of the DO, typically an analog control point. In addition to the 1PPS signal, typical GPSDOs provide a 10-MHz signal for use in a reference frequency for lab equipment, synthesizer reference, etc.

The need to measure the phase difference of the two 1 PPS signals creates the most difficulty in such a design. The phase detector must operate on 1 Hz signals and provide phase difference information typically below a few nano-seconds. Typically, a TIC type method is employed that is implemented in FPGA or similar programmable logic.

The final performance of such systems depends greatly on the design of the PLL system. From the start they are using the 1 PPS signal from a L1 receiver in clock mode. As we now know, these signals phase errors are tied to Code Phase measurements. They are noisy, and long averaging times in the PLL are needed to reduce this noise. The choice of the averaging interval will depend on the stability of the DO used (i.e., Allan Variance). Higher-grade clocks can accept longer averaging times than lower grade clocks. If the averaging time is not chosen correctly, the performance of the final DO output will be compromised.

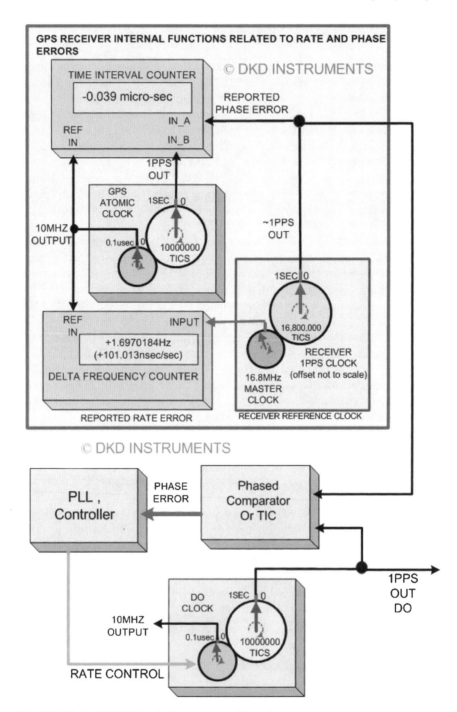

Fig. 8.10 Typical GPSDO hooked up to our model receiver

8.7   A Rate Corrected DDS 5/10 MHz Reference Based on Any 10 MHz Clock

The need for GPSDO's type subsystems could possibly go away if every GPS receiver was able to afford a high-grade master oscillator. But until a small, low-cost atomic clock is developed, such systems will continue to be needed. If we look at the block diagram of our GPSDO (Fig. 8.10) and compare to it to Fig. 8.1, we see that the TIC function is replicated *outside* the receiver for use by the GPSDO PLL system.

This replication of this measurement system is due to the fact that the DO is not providing the receiver's master clock signal. If we can make the DO signal drive the receiver's Master Clock, then we do not need to build our own TIC as was done in for the GPSDO discussed above. *Instead we can use the receiver reported rate and phase errors, which now are referenced directly to the DO.* The precision and resolution of these reported errors is (should be) very high. It is quite difficult to match them with externally constructed circuits.

Figure 8.11 shows such a system that uses a DDS-based DO to provide not only the output 5/10 MHz output signal but also the receiver's Master Clock signal. This architecture allows the GPSDO to use the measurements of Receiver Clock rate and Phase error directly with no external TIC circuitry needed. In the case shown here, two 48-bit DDSs are used, one to generate the master clock at 16.8 MHz and one to generate the 5-MHz output. In receivers where the master clock is 10 MHz, See Zarlink Chap. 9, the DDS for that clock can be omitted with the loss of rate and phase nulling for the receiver clock. If rate and phase nulling of the master clock for use by the receiver is not of interest, then the DDS for that function could also be replaced by a PLL operating on a VCXO.

The design presented here is primarily a rate only corrected output at 5 or10 MHz. The reason is that the high Q crystal filter on the 5 MHz output will introduce significant phase errors that can vary with temperature. The crystal filter is needed to remove close in spurious present on the DDS output.

If a divider were inserted on the produced 5 MHz so as to produce a 1-PPS signal, it would be found to have a phase offset and a wandering phase (with temperature) with respect to the 1 PPS from the receiver. This is an example of syntony as opposed to synchrony. If a low Q filter for the 5 MHz can be tolerated, the phase wanderings with temperature will become significantly smaller, perhaps small enough to where the 5 MHz becomes synchronous to the 1 PPS from the receiver, at least within the accuracy of receiver 1 PPS variation.

The master clock is now any 10 MHz clock signal that has enough stability and accuracy for the receiver to properly operate. To obtain a low rate error 5-MHz output, we simply program the DDS with a rate that accounts for the receiver reported rate error. This is possible since the GPS receiver is providing a rate error measurement of its master clock @16.8 MHz, which is a known rate and phase offset from the supplied 10 MHz clock. In short, we can think of the receiver reported rate and phase errors as being referenced to the 10 MHz Clock.

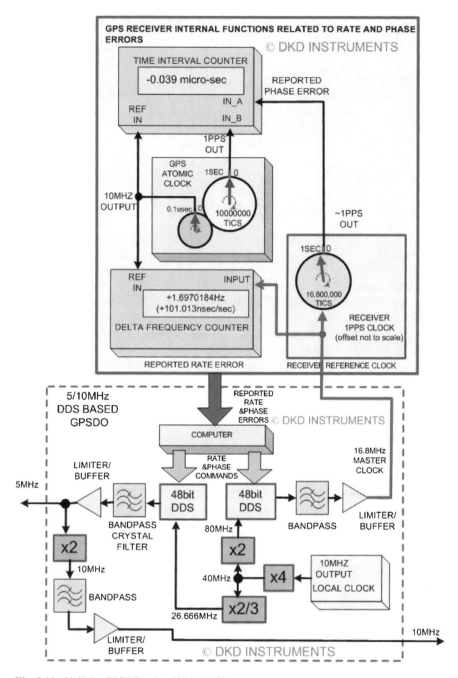

Fig. 8.11 5/10Mhz GPSDO using 48 bit DDS's

In order to produce a clean 5 MHz output from a DDS, we need to provide a clock frequency that is higher than 5 MHz by at least a factor of 3 and that is not an integer multiple of 5 MHz. DDS outputs near integer multiples of the DDS clock (i.e., Commensurate) have spurious frequency components that are very close to the desired output frequency. Multiplying the 10 MHz clock by 4 followed by a multiply/divide by 2/3 results in a DDS clock frequency of 26.6666 MHz. This DDS clock frequency is roughly five times that of desired output rate and is non-commensurate with 5 MHz.

To produce the 10-MHz output, the 5-MHz signal is doubled using a diode-based doubler with typically greater than 40 dB of fundamental rejection. A low Q, double-tuned bandpass at 10 MHz does the final spectrum clean up to produce the rate corrected 10 MHz output.

This design can also be used to drive the receivers 16.8 MHz rate error toward zero if desired, i.e., reported rate errors near zero. Doing so means that rate errors for the 5 MHz will need to reflect this change in reported rate error by backing out the rate error at 10-MHz input. The all-digital, high precision of the design makes this reverse type calculation possible. Driving the rate error to near zero on for the 16.8-MHz signal must be done very carefully for fear of upsetting the measurement of those same rate errors! Leaving the 16.8-MHz rate errors uncorrected has advantages when viewed from a measurement perspective, which is *do not disturb what you are measuring*.

The rate control register of the DDS must have enough bits to provide the rate resolution needed. As DDS clock rate rises, the number of bits needed will increase. For a DDS clock frequency of 80 MHz, at least 36 bits should be available in the DDS. A DDS with 48 bits is significantly more than needed.

The design presented here has multiple advantages over the GPSDO presented earlier;

- All Digital Rate and Phase Control Points.
- Can use a wide variety of 10 MHz Oscillator Clocks as a master reference.
- Uses Receiver-provided measurements of Rate and Phase Error.
- No external TIC circuitry needed.
- Can be used to drive receiver-reported rate error toward zero. In such a case 1 PPS Signal from receiver may have increased fidelity.
- The control loop can easily adapt to the stability of the 10-MHz reference used

For the case where the receiver's clock rate remains uncorrected, i.e., its dedicated DDS output rate control is static and will contain the rate error ε, but we wish to correct the 5-MHz output rate, we can use the receiver reported rate error ε directly in the 5-MHz DDS calculations. For this case, ε is reporting the estimate of rate error on the master oscillator.

The output rate of a DDS (or NCO) is:

$$f_{out} = [M \times f_{clk}]/2^N$$

where f_{clk} is the clock rate input to the DDS, N the DDS Phase Register width and M the integer value of the DDS rate input to achieve f_{out} rate.

The reported rate error, ε, of the GPS receiver is also the scaled rate error on the DDS input clock. We can use the reported error to write the *corrected* DDS clock rate as:

$$f_{clk} = f_{clk\_nom} + \varepsilon \times f_{clk\_nom} \text{ where } f_{clk\_nom} \text{ is the } nominalrate \text{ of the DDS clock (wrt GPS)}$$
$$= f_{clk\_nom}(1 + \varepsilon)$$

$$(8.10)$$

We can find the nominal value of M, M_{nom} that will produce exactly the desired output rate when the rate error, ε, is zero as:

$$M_{nom} = [2^N f_{out}]/f_{clk\_nom} \qquad (8.11)$$

where M_{nom} is the integer to obtain f_{out} when $\varepsilon = 0$

For the general case where ε is not zero, that is the imperfect f_{clk} is the input to the DDS, we can find M from;

$$[2^N f_{out}]/f_{clk\_nom} = M(1 + \varepsilon) \qquad (8.12)$$

Now using definition of M_{nom} from (8.10), we express M in terms of ε and M_{nom};

$$M = M_{nom}/(1 + \varepsilon) \text{ which is as expected , as } M = M_{nom} \text{ when } \varepsilon = 0 \qquad (8.13)$$

For the hardware implementation and errors as presented in Fig. 8.11, we have;

$$f_{out} = 5 \text{ MHz}, N = 48, f_{clk\_nom} = 26.6666666666 \text{ MHz}, \varepsilon = +101.013 \text{ ns/s}$$
$$M_{nom} = 2^{48} \times 5 \text{ MHz}/26.6666666666 \text{ MHz (rounded to nearest integer)}$$
$$= 52776558133380$$
$$M = 52776558133380/(1 + 101.013 \times 10^{-9})$$
$$M = 52776552802262$$

with M at this value, the 5 MHz output is now corrected within the limits of the rate error ε estimate. The round off errors using 48 bits are much smaller than precision of ε.

8.7.1 A Mechanical Model of Rate Corrected DDS 5/10 MHz Reference Based on Any 10 MHz Clock

Figure 8.12 shows a gear model of the rate corrected 5-MHz clock where the receiver's clock remains uncorrected as in the above example. The multipliers of Fig. 8.11 are replaced by gear ratios of the same values which creates new clock dials. All the gear-driven dials are in a simple fixed-phase (integer) relationship

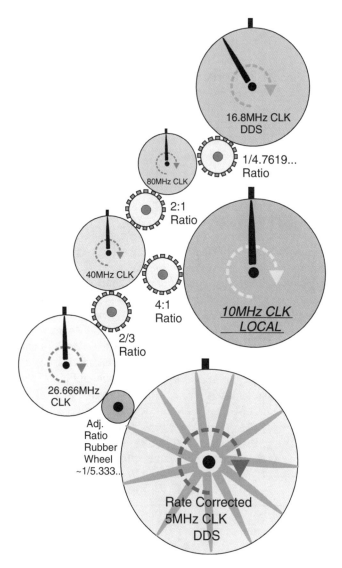

Fig. 8.12 Mechanical Gear Model of Rate Corrected 5 MHz Reference

with the exception of the 16.8-MHz clock dial. This dial is produced by DDS and the ratio is an irrational number. The 16.8-MHz dial phase is highly incommensurate with the driving 80-MHz clock, but it's phase is not random with respect to it. All rates shown are nominal with the 10-MHz local clock as the reference.

The corrected 5-MHz clock dial is modeled as a rubber wheel with approximate ratio of 1/5.333333333[1+ ε]. The rubber wheel can slip so the phase of the 5-MHz dial is random with respect to the driving 26.66666666-MHz clock dial. If we continually adjust the ratio of the rubber wheel using the rate error information supplied by the GPS receiver, the average rate of the 5-MHz clock dial will follow GPS clock rate.

8.8   Receiver Delays in GPS Time Transfer

One of the overlooked delays in GPS Time Transfer, specifically 1PPS output phase error w.r.t GPS Time, is what is called Receiver Delay. In order to explain just how and why these delays occur let us take a step back and form a simple model of time transfer.

In Fig. 8.13 we see two identical clocks on the top of two mountains, a transmitter clock and a receiver clock. Both clocks display zero to one second of time. We wish to make the two clocks read the same time (or have zero phase error w.r.t one another). We will assume that these two clocks run at exactly the same rate. This will allow a one time set of phase and thereafter the clocks will read the same time.

The transmit clock sends out a light pulse whenever it passes its zero timing mark. The receive clock has an arming button that when pressed allows a received light pulse to set its phase or time to read zero. We will call this event a reset. We will assume a zero delay for now between the reception of the light pulse and the setting of resetting of the receive clock to zero phase.

In Fig. 8.13, we see the results at the instant the receive clock is reset. As can be seen, the receive clock had a phase error but now it appears to be reading the same time as transmit clock? Not quite but almost.

As we see in Fig. 8.14, there is a phase offset, DT, in the receive clock w.r.t. the transmit clock. The offset DT is just the time it takes for light to travel from transmit clock to the receive clock. If we know the distance X, we can compute the delay time (DT = X/Speed of Light) and advance the received light flash generated signal to compensate for it. The result of our advance is shown in Fig. 8.15.

This is of course what a GPS receiver is doing with its 1-PPS signal: removing the path delay time from the SV signal resulting in advancing the clock. In the case here and in GPS time transfer, the accuracy of the phase set operation depends on knowing the path delay. But there is also a typically small but significant delay that is not part of the path delay and that is what is often called Receiver Delay.

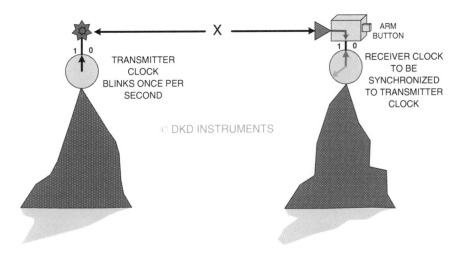

Fig. 8.13   Simple time transfer between two distant clocks

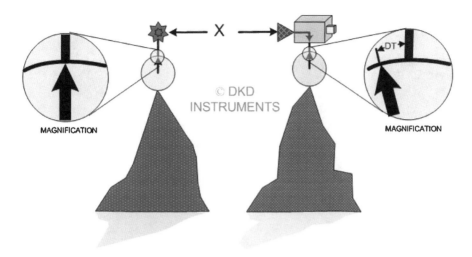

Fig. 8.14 Path delay

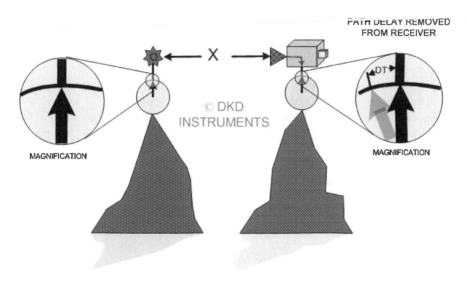

Fig. 8.15 Remove path delay by advancing received signal by DT

In Fig. 8.16, we have introduced a small delay term, RD, at the receive clock. This delay is due to typical hardware delays that occur in any receiver. They can multiple sources, LNA delay, filter delay, cables, etc. We lump them all together (for now) and call them RD.

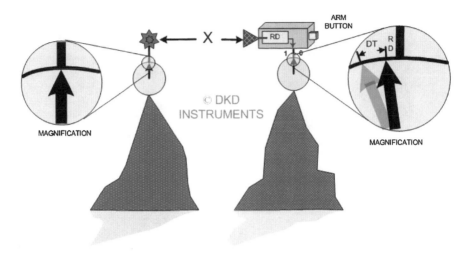

Fig. 8.16 Uncorrected RD delay at reset

As we see in Fig. 8.16, the reset has occurred and the receiver advances the signal to compensate for path delay but we still see a small phase offset due to the delay term RD. We can of course advance the receiver signal as before to compensate for this term. But that requires knowledge of the magnitude of RD.

Unlike path delay, which GPS receivers can calculate from knowledge of receiver position and SV position, receiver delays typically must be measured and entered into the receiver for the particular installation and receiver. In other words, a single GPS receiver on its own cannot determine RD.

Many modern receivers have such a data entry in there software and they should apply this information as to the received signals. Some do not. Also the term RD as discussed here is an amalgamation of many delay terms. Often the manufacturers of the receivers have already compensated for delays *inside* the receiver such as LNA, filters, DSP propagation, etc. But there is one delay they typically do not account for and that is the cable delay from the antenna to the receiver as this varies typically by each installation.

In the modern use of the term Receiver Delay, it refers exclusively to the delay internal to the receiver itself and not external coax feeds, etc. In addition, these receivers should have compensated for the internal receiver delay in their solutions and output corrected phase w.r.t. GPS Time or phase.

8.9 Antenna Phase Center

A related concept to receiver delay is antenna phase center. It is related because at its core is the signal summing and common delay phenomenon of the summed signals associated with typical single element, GPS L1 antennas.

In Fig. 8.17, we see a typical receiver system composed of four GPS SV signals being received by a single, common L1 antenna. We will assume our antenna has

Chapter 9
The Zarlink 12-Channel GPS Receiver

In this chapter, we take a brief look at the two-chip Zarlink GPS receiver, Zarlink part numbers GP2015 and GP2021. The GP2015 performs the RF downconversion and the GP2021 does the baseband processing. This receiver is commercially available and has software that is open Source, see Clifford Kelly's website http://www.earthlink.net/~cwkelly. Our discussion will concentrate on how the hardware of this product works. The GP2021 can also be used to receive the GLONASS signal, here we only discuss the GPS use of this chip.

9.1 The Zarlink GP2015 RF Downconverter

This device converts the GPS signal at 1,575.42 MHz to an analog IF signal at 4.309 MHz. In addition, a digital IF at 1.405 MHz is provided as sign and magnitude bits. Figure 9.1 shows a block diagram of the GP2015.

9.1.1 Triple Conversion to 4.039 MHz IF

Three mixer operations are used to downconvert the GPS signal to the analog 4.039 MHz IF. This analog IF output is not used by the GP2021 baseband processor, rather the GP2021 uses the digital IF @1.405 MHz. Each conversion stage has a corresponding LO which is synthesized on the chip. This includes LO1, which is a phase-locked oscillator. The other oscillators are derived using frequency division of LO1. The reference frequency for the LO1 synthesizer is 10 MHz. A TCXO oscillator would typically provide the 10 MHz reference.

The IF bandpass filters are external elements that must be connected to the GP2015 with the exception of the 4.309 filter, it is on the chip. These can be SAW

D. Doberstein, *Fundamentals of GPS Receivers: A Hardware Approach*,
DOI 10.1007/978-1-4614-0409-5_9, © Springer Science+Business Media, LLC 2012

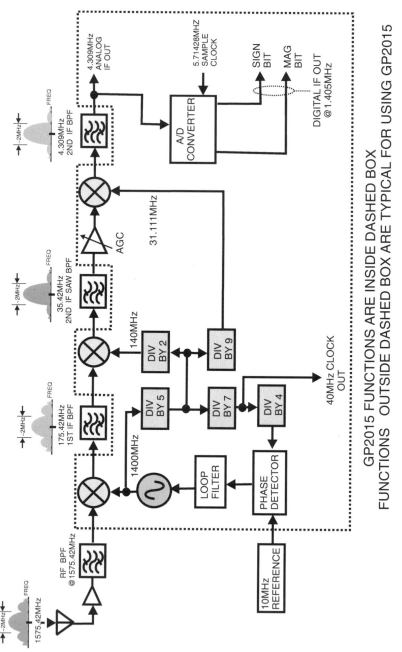

Fig. 9.1 Zarlink GP2015 GPS RF downconverter (simplified)

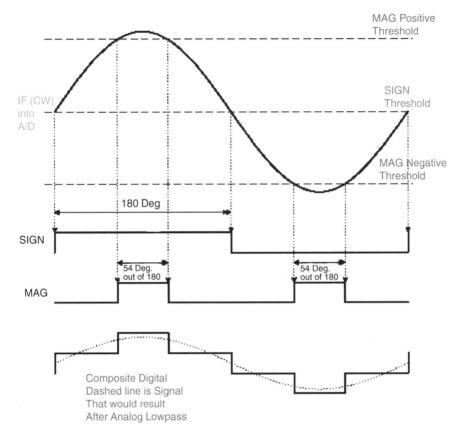

Fig. 9.2 Sign and magnitude representation of sinewave

filters or discrete filters. Typically, a SAW filter is used for the second IF at 35.42 MHz. This filter should have a 3 dB bandwidth of approximately 2 MHz as the signal is wideband through out the IF conversion process.

9.1.2 Digital Sampling Creates IF@ 1.405 MHz

The 4.039 MHz third IF is digitally sampled to create a fourth IF frequency at 1.405 MHz. This IF is a digital IF and does not have an analog output. In particular, it is a rather messy digital waveform at the output of the GP2015. The Digital output is available as a SIGN and MAGNITUDE bit. This approximates a sinewave as shown in Fig. 9.2. The low number of bits used to capture the GPS signal may seem like a crude approximation. But it is important to remember that the GPS signal contains only Phase Modulation. Phase information is contained only in the zero crossings of the signal and not in its amplitude. This allows a low number of bits to be used to represent the GPS Signal.

The sampling process performed at 5.71428 MHz creates the 1.405 MHz digital IF. In many respects, the Sampling process is acting like a mixer. By sampling the 4.309 MHz IF at 5.71428 MHz, the spectrum is repeated at various spots in the frequency domain. Since no filtering is provided by the GP2015, almost all of these possible spectral locations are available to process by subsequent digital methods.

For the GP2015/2021 chip set, the difference spectrum at 1.405 MHz is chosen for processing. Centered here is the GPS signal in its Uncorrelated, wideband form with a bandwidth set by the filter at 4.309 MHz (approx. 2 MHz). This is essentially the lower side band product from the "mixing" of the 4.309 MHz IF and the 5.7128 MHz sample clock.

9.1.3   GP2015/GP2021 Clock Signals and Complex Mode

Two important clock signals are used by the GP2015 and by the GP2021. First is the sample clock at 5.71428 MHz. The second clock signal is the 40 MHz clock derived from division of LO1@1,400 MHz. To be absolutely clear, all clock signals and LOs are derived from the common reference at 10 MHz through division or phase-lock processes. In this respect, the GP2015/2021 pair form a coherent receiver chip set.

The 5.71428 MHz sample clock is generated by the GP2021 by division of the applied 40 MHz-clock signal. It uses this clock for its own digital processing methods and provides it for use by the GP2015 to sample the IF at 4.309 MHz.

The 40 MHz-clock signal is used by the GP2021 for generating a multitude of internal timing signals. A short list of these is TIC, the internal DCO clocks, MICRO_CLK, etc. The 40 MHz clock is provided as non-inverted (CLK_T) and inverted (CLK_I). It is also possible to use the GP2015/2021 in a complex mode. In this configuration, some clock frequencies are changed from those presented here. In this discussion, we will focus on the REAL mode of operation and that set of clock frequencies. For Complex operation, the reader should consult the Zarlink data sheets.

9.1.4   The TIC Signal

The TIC signal performs the same function that the SNAP_SHOT signal does in the GPS100SC receiver. It is normally set to ~1/10 of a second. Just as in the GPS100SC, the TIC captures the state of the SV replica clocks present in the GP2021. In the GP2021 that means 12 SV replica clocks are captured every TIC signal. The structure of the SV replica clocks is slightly different compared to the GPS100SC. The interpretation of the data is exactly the same, it is an estimate of the uncorrected "time sent" information for each SV tracked.

9.2   ZarLink GP2021 12-Channel Baseband Processor

The GP2021 is a 12-channel baseband processor. It can do C/A code correlation and Doppler tracking on 12 SV signals simultaneously. To do the equivalent operation using the GPS100SC receiver would take 12 replicas of the signal processing from the second IF onward. Obviously in terms of integration, the GP2021 is a marvel. Figure 9.3 shows a block diagram of the GP2021. This integration is accomplished through using all digital techniques. Interestingly, the GP2021 is not a complete baseband receiver implementation. It needs a computer to complete the control loops associated with C/A code Scan and track as well as Doppler Scan/Track, Data demodulation, etc. What the GP2021 provides the host computer is all the control points and data gathering registers to do Code Lock, Doppler Lock, SV clock regeneration, etc. in a completely digital fashion.

Our discussion will focus on how the GP2021 obtains C/A code Lock, Doppler Lock, and 50 Hz Data demodulation.

9.2.1   Single-Channel Block Diagram

The 12 channels of the GP2021 are identical when used in the REAL mode. Figure 9.4 shows a block diagram of a single channel. The 40 MHz clock mode is selected (REAL Mode).

In many ways, the processing here is similar to the GPS100SC receiver. In the block diagram, the various +/− numbers indicate the possible states the digital signal can achieve at that point. Let us follow the signal as it is processed in the GP2021.

9.2.2   Doppler Offset Removal

At the first mixer, the Doppler offset is striped away. This is accomplished by mixing the incoming digital IF at 1.405 MHz that has +/− Doppler on it with a LO that has this exact same frequency. This creates a zero-frequency signal (Baseband) for subsequent processing. Instead of a VCXO as we used in the GPS100SC receiver, a DCO (Digitally Controlled Oscillator) is used here for this purpose. A DCO is just a DDS without the DAC portion. The Doppler DCO in the GP2021 has a frequency resolution of about 42 mHz (40 MHz Clock). By comparison the GPS100SC has a Doppler resolution of approximately 10 kHz/1,024 ~ 9.7 Hz. The extremely fine Doppler resolution is needed to precisely match the incoming phase and frequency of the incoming GPS signal so as to mix it as nearly as possible to true baseband (Approximately Zero Phase and Frequency Offset).

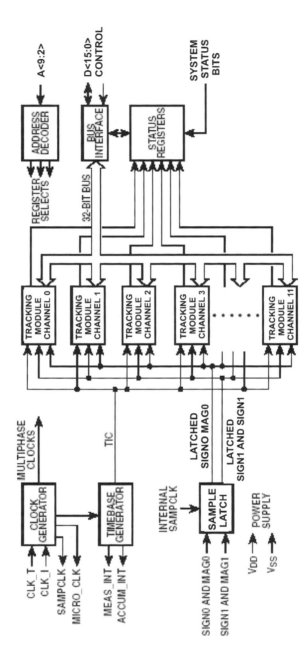

Fig. 9.3 Block diagram of GP2021 12-channel baseband receiver

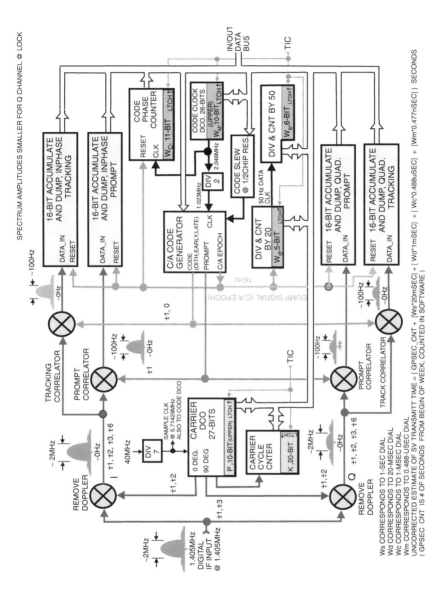

Fig. 9.4 Block diagram of a single channel of GP2021 (spectrum shown in lock mode)

The GPS100SC used a single mixer for correlation and a single mixer for Doppler strip off. The GP2021 uses In-phase and Quadrature (I and Q) processing for Doppler removal. This I and Q processing is also used by the C/A code-loop processing. The purpose here is to not only to facilitate the remove the Doppler but also to demodulate the 50 Hz data at the same time. By properly using the information provided by the I and Q arms, the Doppler loop can be closed. The host computer must close the Doppler loop in software.

The computer uses the data gathered from the I/Q arms to dynamically correct the DCO frequency. It does this by reading the 16-bit accumulator information. Exactly how these accumulators work is covered below. The control loop must constantly measure and correct the DCO so as to keep it nearly equal in frequency and phase to the received IF at 1.405 MHz. This is of course exactly as was done in the GPS100SC receiver. The difference is the coherent design and frequency precision of the GP2015/2021 allows this to be done without the frequency counter used in the GPS100SC design. The particulars of the Doppler control loop are discussed below.

9.2.3   C/A Code Sliding Correlators

After the two Quadrature mixers used to remove Doppler offset, four mixers are encountered. These are all used for C/A code correlation. Each one of these mixers is used as a sliding correlator. In the GPS100SC, we only had one correlator used both for Code Scan and Track. The additional correlators are used to facilitate the capture process of the C/A code. They also provide a PROMPT channel free of induced modulation used to drive the code-tracking loop.

The two correlators on the INPHASE arm are used for tracking the C/A code and also aid in the capture process. The two correlators in the QUADRATURE arm normally are used only during the signal-acquisition process. The reason is simple. Once Doppler has been properly tracked the signal level in the QUADRATURE arm is very low. Just the opposite is true for the INPHASE arm. Once Code and Doppler processes are in track mode the INPHASE arm will have most of the signal energy.

Like the Doppler process the C/A code correlators output is obtained by reading the 16-Bit accumulators. Unlike the GPS100SC where there were separate structures for the Doppler and Code tracking loop information, these two loops share common structures in the GP2015/2021 to accomplish these tasks. In particular, the 16-Bit accumulators provide information for both the Code and Doppler tracking loops.

9.2.4   C/A Code-Clock Generator

The code-clock generator in the GPS100SC was a simple shift register used to divide down and modulate the TCXO referenced frequency at 20.46 MHz. In the GP2021, the code clock is a DCO very similar to that used in the Doppler loop. This oscillator has extremely fine frequency and phase resolution. It creates the C/A code by direct synthesis using a clock signal derived from the 40 MHz CLK_T signal, see above. Digital commands can change the frequency of the code-clock generation. Also, the phase of the code clock can be set to a particular value after a reset event.

9.2.5   Prompt Channel, Early, Late, and Dither Codes

The GPS100SC used the Tau-Dither code tracking method. In Tau-Dither, one correlator is used for both tracking and the channel that will be used to demodulate the 50 Hz data. In the GP2021, an extra correlator is available for the PROMPT channel. The PROMPT channel usually is not used in the C/A code acquisition or track process. Rather, it is the channel where the demodulated data and Doppler Information is extracted.

The GP2021 has several modes available for the code-tracking loop. The most unusual is the case where a C/A code replica is created that has both Early and Late codes combined into a single signal EARLY-MINUS-LATE signal. Additionally, the GP2021 can provide a DITHERED code to one correlator in each arm. In this mode, it resembles the GPS100SC the most. The difference is that there is a PROMPT correlator with the C/A code at the midpoint between the two dithered codes. The DITHERD code is switched at a rate of 50 Hz. In the GPS100SC, we used 166 Hz.

9.2.6   C/A Code Scanning, Slewing

In the GPS100SC, the C/A code was scanned by the same mechanism used in the track function, the Code-Clock Modulator. In the GP2021, things are more compli-cated. A method to advance/retard the C/A in ½ chip steps is provided for code scanning or *slewing*. This allows the scanning process to get the C/A code within ½ chip of alignment. If this triggers a correlation detection event, the remainder of the code alignment is done by adjustments to code-clock frequency via the code DCO. The Code DCO is not adjustable in phase (except at reset) so the remaining code misalignment is corrected by adjustments to the DCO frequency register. This changes the C/A code-clock rate, which moves the replica code with respect to the received code.

9.2.7   Code-Phase Counter and Code-Clock Phase

Just as in the GPS100SC, a method is needed to record the state of the replica C/A code generator. The GP2021 has two places to retrieve code-phase information.

The code-phase state counter is done more or less the same as done in the GPS100SC receiver except it records to a ½ chip resolution. This is a linear counter whose contents reflect the Code State in ½ chip increments. This is latched by TIC (typically, every 1/10 s) and can be read by the host computer. This has a time resolution of one C/A code chip or about ½ μs. This is of course is the 1 ms dial of the replica clock with twice as many increments (2,046 total) compared to the GPS100SC version (1,023 total).

Additionally, the phase register of the code-clock DCO can be recorded at the TIC time signal. The DCO phase register resolution is very high at 1/2,048 of a chip. This works out to be a time resolution of about 0.5 ns. In the GPS100SC receiver, we had 20 increments on the 0.977 μs dial. The GP2021 has 1,024 increments on a 0.488 μs dial! Figure 9.5 shows the GP2021 replica clock. In theory, the GP2021 can resolve down to about 0.15 m. But jitter in the C/A code loop and other uncertainties insure this accuracy will not be achieved in the range measurement with just C/A code measurements, see Chap. 8.

9.3   The 16-Bit Accumulators

Looking at the overall structure of the GP2021, we see many similarities to the GPS100SC receiver. The two largest differences are the use of digital data accumulators and the I/Q processing. We have already spoken of both of these methods. Now, we will focus on the details Accumulator and later the I/Q process as it is used for the Doppler Loop.

9.3.1   How Do the 16-Bit Accumulators Work?

First, let us be sure what the accumulators do. The accumulators in the GP2021 are acting as adders. Every sample clock (@5.71428 MHz) the input data is added to existing contents of the accumulator. This process continues for 1 ms. At the end of 1 ms, the accumulator contents are latched for the processor to read. Then the accumulator is reset to zero (DUMPED) and the process starts all over.

If we assume the maximum input of $+ 6$ is applied to an accumulator, then we can calculate the maximum value that would accrue after 1 ms with a sample clock of 5.71428 MHz.

$$\text{Number of samples} = 5.71428 \, \text{MHz}/1 \, \text{kHz}$$
$$\Rightarrow 5,714 \text{ samples accumulated in 1 ms} \qquad (9.1)$$

If $+ 6$ is applied for all 5,714 samples, the value accrued would be $+ 34,285$. This is just slightly larger than can be held in a signed 16-bit register, which has a

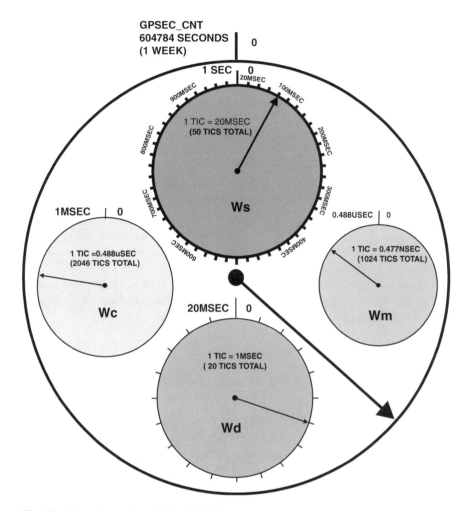

Fig. 9.5 SV replica clock model for GP2021

capacity of ±32,768. In use, the probability that the + 6 would be applied over one 1 ms would be zero. We can say the register is big enough.

Now that we know the details of operation, we need to look at the function this accumulation operation performs.

9.3.2   The DUMP Signal

The digital accumulators are reset to zero by this signal as we just learned. This signal corresponds exactly to the C/A epoch signal, which occurs at a nominal rate of 1 kHz. Therefore, the DUMP signals for each channel are different and will not be synchronous with each other, unless two channels are set to receive the same SV signal. The DUMP signal not only resets the accumulator to zero but also resets the Code-Phase Counter to zero and Resets the Code Generator to its "initial" state.

By tying the DUMP signal to the 1 kHz C/A epoch, the effect is to synchronize the integration period to the timing of the incoming GPS signal. Of course, this statement is only true once signal TRACK is achieved. Assuming tracking is achieved, the integrators will start at the beginning of C/A code cycle and reset at its end. Also, there will be exactly 20 integration periods for each data bit.

9.3.3   Digital Accumulators as Integrators

The action of digital accumulation is closely related to integration. Integration is inherently a smoothing or averaging operation when we think of its effects on an applied signal. Knowing this, we can say that the accumulators perform signal smoothing and averaging. But we can go further and state that digital integrators that are cleared (or DUMPED) at regular intervals of time act like a lowpass filter with a cutoff frequency of about half the DUMP frequency.

9.3.4   Approximating a Digital Accumulator
as Analog Lowpass Filter

To understand the lowpass approximation, examine Fig. 9.6. This figure shows the contents of the accumulator as a function of time for two sinewave inputs. For this example, assume that the sample rate is lower than actually used so as to show the "steps" in the accumulation and that the input resolution (in bits) is fairly high.

In the first case, the sinewave has a frequency of 100 Hz. For convenience, the DUMP period is shown synchronous with the sinewave and starts at the beginning of the sinewave for both input examples of Fig. 9.6. We see the accumulator contents rising and then being cleared every 1 ms. Since there are ten 1 ms intervals in a 100 Hz sinewave, the accumulator is dumped ten times.

Looking at the resulting accumulator output one could say it is not doing much smoothing, in fact it is a saw tooth-like waveform and rough. But it is also apparent that the peak value of accumulator contents are following the amplitude of the sinewave. So in some sense the accumulator contents can be seen as a sinewave.

Now let us look at the case where a sinewave of exactly 1 kHz is applied to the accumulator input. Now what happens is the accumulator contents rise to a maximum at the point where the sinewave goes from positive to negative. Then the contents of the accumulator start decreasing as we are now subtracting values, as the signal is negative in value. At the end of the applied 1 kHz sinewave period, the accumulator contents are zero.

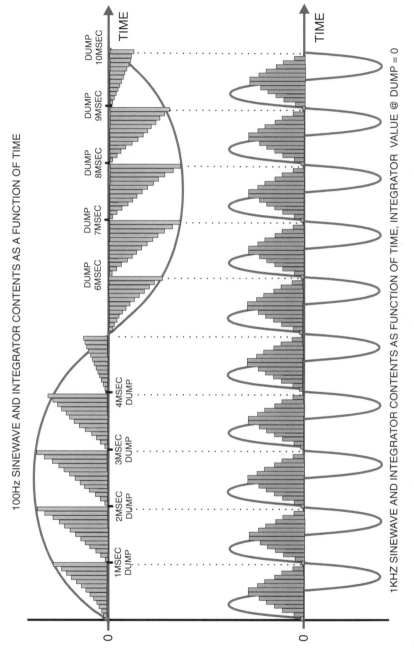

Fig. 9.6 Accumulator contents as function of time with sinewave-applied integration interval (sample clock) not to scale

Now we can say that for a digital accumulator, with a DUMP cycle of 1 ms and input sinewaves with frequencies above 1 kHz, the *average* results in accumulator contents are zero. And we can say that for sinewaves with frequencies below about 500 Hz, the *average* result in the accumulator follows the amplitude of the applied sinewave. It should be clear now that the digital accumulators in the GP2021 are acting like lowpass filters with a cutoff frequency of about 500 Hz.

9.4   An Analog Model of the Doppler Loop

Now that we have an analog model of the digital accumulation process, we can form an analog model of what is happening in the Doppler Loop of the GP2021. Figure 9.7 shows our model. For this model, we have eliminated the code correlation mixers and will assume that the C/A code is removed from the signal. We will also assume that the 50 Hz data modulation is still present. The Doppler DCO will be modeled as an analog VCO.

This model will enable us to approximate what the outputs of the accumulators will be for the Doppler tracking Loop and to some extent the Code tracking loop.

9.4.1   Assume VCO Is Exactly Correct in Phase and Frequency

In this situation, the VCO is exactly in phase with the received signal. In other words, the Doppler track system has "locked on" to the applied signal. Figure 9.7 illustrates the model with conditions as just described.

Looking at the Quadrature arm of the processing the result of the multiplication of the VCO with the received GPS signal is the product of two sinewaves at the same frequency that have a 90-degree phase shift between them. The lowpass filter will average this waveform and produce a signal of approximately zero volts. In this case, we can say that the Quadrature arm of the Doppler signal processing has nearly zero energy in it when the VCO is locked onto the incoming GPS signal.

If we look at the In-phase arm we see two possibilities. The first is that the 50 Hz data bit is a "1." In this case, the incoming sinewave and sinewave from the VCO are exactly lined up as shown in Fig. 9.7. The second possibility is that the data bit is a "0," or a negative voltage level in our analog model. This results from the input being 180° out of phase with respect to the VCO output. The negative voltage would be recorded as negative integer in the digital accumulator. These negative outputs are of course due to the BI-Phase modulation used for the 50 Hz data. In both cases, the mixer multiplies the two sinewaves together and then the result is lowpassed.

The output of the lowpass for the case where the data bit was 1 is essentially a DC level corresponding to the data bit "1." As we just discussed, when the data bit is zero the output of the lowpass filter would be a negative voltage (negative

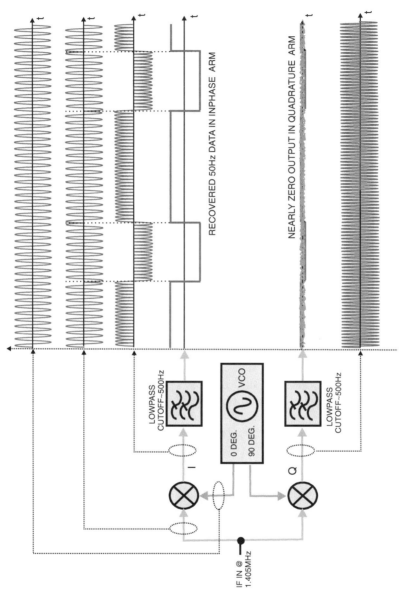

Fig. 9.7   Analog model of I/Q processing

integer as read from accumulator). By reading the accumulator contents of the Inphase arm, the host computer can recover the 50 Hz data from the GPS SV being tracked.

Now, it is clear that the In-phase arm of our model will demodulate the 50 Hz data as long as we keep zero error on the VCO in *phase and frequency* with respect to the incoming signal from the SV. It is not trivial to get from a situation where there are large frequency and phase errors to one where they are quite small, as we shall see.

9.4.2   Adjustable Baseband Bandwidth for Track

Its no coincidence that the accumulators are dumped every 1 ms. As we now know, this results in an equivalent lowpass operation on the mixer output of the Quadrature and Inphase arms of the GP2021. Baseband lowpass filters of 500 Hz bandwidth are equivalent to 1 kHz wide bandpass filters if they are shifted in the frequency domain away from DC.

1 kHz is the bandwidth used in the GPS100SC receiver at the 10.7 MHz crystal IF filter. A 1 kHz bandwidth Bandpass filter or equivalently 500 Hz for baseband processing is a good bandwidth for capture of the GPS signal. But it is not the optimum bandwidth once the GPS signal is captured and tracked in Code and Doppler.

Once tracking is achieved it is quite desirable to "tighten up" the receiver bandwidth. With the GPS100SC, this not possible as the IF bandwidth is fixed. With the GP2021, it is possible to reduce the bandwidth of the equivalent lowpass filters implemented by the 16-Bit accumulators. Adding two successive accumulator results together can do this. If this is done, we have halved the bandwidth of the equivalent lowpass filters. This illustrates how easy it is to adjust the bandwidth of the GP2021 signal processing at least at the level of the host-computer processing.

9.4.3   Doppler, Code Scan, and Threshold Detects

As with the GPS100SC receiver, the GP2021 must scan the possible Doppler offsets until the signal falls inside its 500 Hz lowpass filter. Commanding ramp type frequency values to the Doppler DCO does the scan operation. The software monitors the combined signal level in the I and Q arms while scanning Code and Doppler. If a predetermined threshold is exceeded, then C/A Code and Doppler tracking are initiated. This process is essentially identical to that of the GPS100SC. The thresholds are set in software, which is an advantage as it allows them to be adjusted for changing signal conditions if needed.

9.4.4   Doppler Acquisition and Track

Previously, we assumed the Doppler loop was "locked on" to the applied GPS signal. But how can we use our model to explain how the VCO can be commanded to the correct frequency and phase?

This is not an easy question to answer. The methods used to "pull in" the VCO are usually different than those used to command the VCO once track is established. Looking at Fig. 9.8, we can see that when the VCO frequency is significantly off the output of the lowpass filters in each arm are fairly complex signals. And we have simplified them by assuming an analog model and that the C/A code was not present!

Generally speaking, in "lock" the energy in the Inphase arm is at a maximum while the energy in the Quadrature arm is at a minimum. By steering the VCO such that this condition is achieved the phase and frequency of the VCO will move toward synchronism with the applied signal. But this can be problematic.

Another approach is to use the lowpass filter outputs to estimate the angle between the VCO sinewave and the applied GPS signal sinewave. The inverse tangent function can be used to estimate the angle between the two sinewaves. The ratio of the accumulator values in the Inphase and Quadrature arms provide input to the arctangent function. This results in a computed, estimated angle. From the angle information, an error signal is derived and applied to the VCO so as to drive this angle toward zero.

These two methods rely on the host computer to read the accumulated data every 1 ms and properly process it. The most difficult part of this process is transition between SCAN and TRACK particularly for the Doppler loop. We will return to this discussion after a more detailed look at the waveforms of Fig. 9.8.

9.5   Analog Model Approximates Unlocked Output Waveforms

Up to this point, we have only examined waveforms in the locked state. In the locked state, the waveforms are pretty well-behaved and given enough SNR they will stay that way. It is the waveforms associated with the transition from scan or unlock that get messy as we have just discovered. Let us take a closer look at a few samples of them from our analog model as shown in Fig. 9.8.

Figure 9.8 shows some typical lowpass outputs for three different conditions. In each case, it is assumed the channel shown is unlocked or equivalently in a "not tracking" condition. Three cases are examined. Two from PROMPT channel processing and one from Dither channel processing.

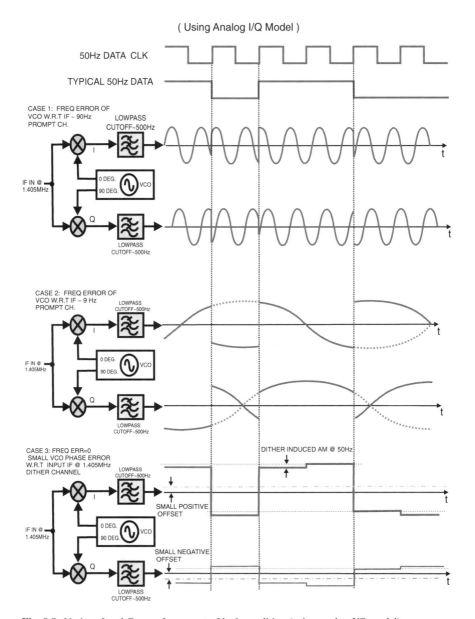

Fig. 9.8   Various I and Q waveforms, out of lock condition (using analog I/Q model)

9.5.1   Case 1 Frequency VCO in Error by 90 Hz

In Case 1, the VCO has a frequency difference of about 90 Hz with respect to the received signal frequency. This frequency difference will manifest itself as a "beat note" or "beat frequency" in the output of the lowpass filters. This is shown in

Fig. 9.8. The remaining 50 Hz data modulation still continues to Bi-phase modulate the beat frequency output. The inphase and Quadrature signals will have approximately equal amplitudes as the phase angle between them is continually changing.

9.5.2   Case 2 Frequency of VCO in Error by 10 Hz

In case 2, the VCO is very close to the frequency of the input, it is only about 10 Hz off. Still the outputs are nearly equal in amplitude. With a 50 Hz data stream and a 10 Hz "beat note," the resulting sinewaves in the I and Q arms are chopped up by the data. The dashed lines on the output waveforms show the parts of the sinewaves as they would be with no data modulation. The waveforms shown are from the PROMPT channel.

If the DITHERED mode was ON and we looked at that channel and its correlator, we would see the amplitude of the 10 Hz beat note change every 20 ms as the dither modulation is at 50 Hz. Until the DATA clock is synced to the incoming data, the Dither AM most likely will not be synchronized with incoming DATA bits. This would only happen if by chance the replica of the data clock is in sync with incoming data clock.

9.5.3   Case 3 VCO Frequency Error is Zero,
Small Residual Phase Error

In this case, the beat note is a DC value. That is the difference frequency between VCO and input signal is 0 Hz. Instead a small phase error is assumed. The phase error manifests itself as DC voltage, or offset, on both I and Q arms. If this small angle is driven to zero, the Quadrature arm signal level will also be near zero while the Inphase arm signal level will be a maximum.

At this point, closing the Doppler Loop is not too hard. If the angle between the VCO sinewave and the incoming sinewave is continually pushed toward zero, the frequency and phase error of the VCO will also be driven toward zero. It is this case that is most often discussed and analyzed. The reason is it is simple and fairly linear. The problem is getting the VCO frequency error small enough to where this case approximately applies so we can lock onto the incoming signal.

This last case shows how the I and Q processing can be used to derive an error signal that when properly used can "finish up" the acquisition process. What is needed is a way to determine which way to move the VCO frequency when the frequency error between VCO and input signal is large, see case 1 above. A method to get a frequency error signal, or discriminator, from I and Q processing is present below.

9.5.4   *Getting Code Lock Using I and Q Data*

The error signals associated with the Code-Tracking loop are by the nature AM signals. This true for all the Code-tracking methods available for use in the GP2021. Because of this, it is possible to extract C/A code tracking information with Doppler errors present. The C/A code tracking information can be extracted from the I and Q processing even in the case where there is a high beat note. The I and Q dump samples taken at every 1 ms are simply squared, summed, and then the square root operation is done. This gives an approximation of the incoming signal amplitude regardless of the presence of a beat note or not. The amplitude data is then processed to extract the error information for C/A code tracking. In short, it is not necessary to achieve Doppler lock to achieve Code lock in the GP2021. This of course is the same for GPS100SC receiver.

In the GP2021, we have no explicit frequency counter as we have in the GPS100SC. This makes getting Doppler lock harder to achieve than Code Lock in the GP2021.

9.6   Getting Frequency Discriminator Information from I/Q Processing

As we have just been discussing, what is lacking in the GP2021 is an overt frequency discriminator as we had in the GPS100SC receiver. One does exists in the I and Q processing of the GP2021 but it is subtle and must be implemented in software.

We need to think about what happens when the VCO is off a bit in our simple analog I and Q model. We can define the difference frequency as the incoming frequency minus the VCO frequency. This is of course the beat note frequency. The VCO can be above or below the incoming IF @ 1.405 MHz. If it is below then we get a positive difference frequency. If it is below we get a negative difference frequency. In the positive case the Quadrature output will *lag* the Inphase output. If the difference frequency is negative the Quadrature arm will *lead* the Inphase arm. Thus frequency discrimination information does exists in the I and Q processing in the form of the lead/lag relationship of the I and Q arms.

Figure 9.9 shows our simple model with a 100 Hz-frequency difference. The mathematics shown illustrates what has just been explained. In particular, the sine function is an odd function and passes a negative argument as a change in amplitude (to negative). A change from a positive to negative amplitude is the same as 180° phase shift for sinewaves. This is of course the sign change that occurs when the VCO frequency is greater than the incoming frequency and results in 180-phase shift in the Q arm signal. The Inphase arm is unaffected by this sign change as it is assumed to be the cosine function, which is even. The sign of the Cosine function is unchanged by changes to the sign of its argument.

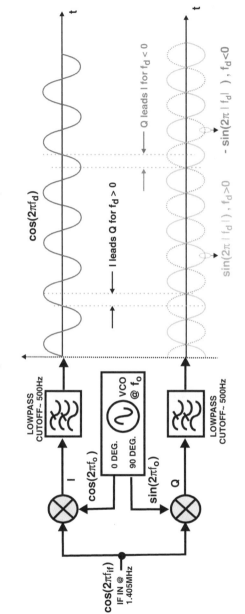

Fig. 9.9 Frequency discrimination using I/Q analog model (assume |*f*d| less than 500 Hz)

If software is written to determine which arm (I or Q) crosses zero first (for the same slope sign) then a determination of the S*ign* of the VCO frequency error can be made. If in addition a software counter is used to count the time elapsed between zero crossings in either the I or the Q arm, an estimate of the Magnitude of the VCO frequency error is obtained. Combining them, we now have an estimate of the Magnitude and Direction to send to the VCO so as to "drive" its error to zero. This course is a Frequency Locked Loop.

Once the FLL gets the VCO error small enough we can switch to a phase based error loop (or PLL) using the derived angle error as outlined above. The FLL is not precise enough to drive the Phase Error of the VCO to zero and unless this is done we cannot demodulate the 50 Hz data or get accurate cycle count data.

9.6.1   Analytic Signal Interpretation of f_d Sign Change

We can interpret the I and Q arm signals as the real and imaginary components of an *Analytic* signal. An Analytic signal is one that cannot be represented as a *Real* waveform or spectra. Rather when it in an actual circuit, it is split into two paths as done here. The mathematical expression for such a signal is:

Analytic signal for I and Q processing model $= \cos(2\pi f_d t) + j \times \sin(2\ \pi f_d t)$

where j is the imaginary number

$$(9.1)$$

In terms of vectors this is a single vector rotating at rate f_d. If it where a Real signal there would be two counter rotating vectors at rotation rate f_d. For the analytic signal, the direction of rotation depends on the sign of f_d. When there is a sign reversal, the rotation direction reverses. In a Real signal this information is lost. In a Real signal (sinewave type), there are always two vectors rotating in opposite directions and even though the change of direction could happen it cannot be discerned because of this.

9.7   Cycle Counting in the GP2021

As we will discuss in Chap. 10, it is possible to estimate the change in range between user and SV by integrating the Doppler information. In the GP2021, a high-resolution counter system is provided for this function. It has a counter that counts integer whole cycles of the Doppler DCO frequency and the ability to read the phase bits of the DCO for very fine resolution. To be clear, this counting operation is only done on the Doppler, or LO_{if}, DCO output frequency. Both of these values are caught at the TIC instant.

In the GP2021, all frequencies are phase locked or divided/multiplied from a common reference signal, 10 MHz source. If this source is an extremely high-quality one, then the counts from the cycle counter will also be a very accurate estimate of the *true* Doppler present on the received GPS signal plus a bias rate equal to the final IF frequency. It must be remembered that the accuracy of the received Doppler information is limited by the accuracy of the 10 MHz reference used. For a more complete discussion of carrier phase measurements and a software method of accumulating integrated Doppler from Carrier phase residual, see Chap. 10.

9.8 Summary

In this chapter, we examined the Zarlink GP2015/2021 chip set. This chip set is a hybrid software/hardware system that can be used to implement GPS receiver. The hardware only does so much, then it is up to the user to write a substantial amount of software for the host processor to complete the receiver. The GP2021 demonstrates how DSP processing replaces some of what was done using analog methods for the GPS100SC receiver. It is hoped with an understanding of the GPS100SC and the GP2015/2021 receiver that the reader can now comfortably move on to more advanced and modern designs.

Chapter 10
Carrier Phase Measurements and Turbo Rogue Receivers

In this chapter, we examine the measurement of Carrier Phase and related topics. Several of the subsections here have to do with analysis inspired by the work of JPL as manifested in the Turbo Rogue receiver architecture. The Turbo Rogue receiver was funded by NASA and executed at JPL. In Chap. 11, the Turbo Receiver is examined in detail.

Although the analysis presented here has the Turbo Rogue as a reference, much of the work is applicable to GPS receivers in general.

10.1   A Mechanical Clock Model of Carrier Phase Range Measurement

We can use the mechanical clock models developed in Book I to help understand the carrier phase measurement process. In Fig. 10.1, we see the mechanical clock model we will use. The distant SV L1 carrier phase dial has a delayed image (delay = Range/Speed of light) at the receiver's reference plane. This delayed L1 Carrier Phase Dial image has an apparent rate to the observer at the receiver reference plane that is above or below the nominal rate of 1,575.42 MHz by exactly the Doppler rate at that instant.

In addition to the delayed image of the SV Carrier Phase dial, the receiver has a local copy of the L1 Carrier Phase dial that has zero rate and phase difference with respect the L1 dial in the SV. By continuous phase comparison of our local copy of the SV L1 carrier dial with delayed image of the received L1 dial, we can make range measurements from the receiver to the SV with extreme resolution (typically around a centimeter).

D. Doberstein, *Fundamentals of GPS Receivers: A Hardware Approach*,
DOI 10.1007/978-1-4614-0409-5_10, © Springer Science+Business Media, LLC 2012

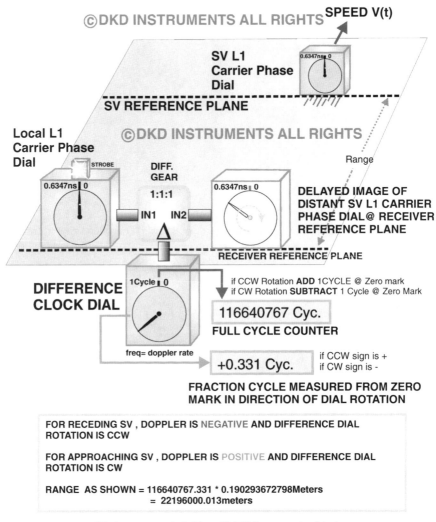

Fig. 10.1 Mechanical model of carrier phase measurement

In order to investigate the phase measurement process, we use a differential gear mechanism to create a new clock dial that is the difference, in both rate and phase, of the Delayed Image of SV L1 dial and the local copy of the L1 dial. This difference dial contains all the information we need to make our Carrier Phase-based measurement.

10.1.1   Observer Sees Pure Doppler and Static Phase Offsets on Difference Clock Dial

To the observer at the receiver reference plane, the difference clock dial will rotate at the Doppler rate established by the relative movement between the receiver and the SV on the LOS. If there was no continuous rate of movement but just small displacements less than a $L1$ wavelength, those small static displacements would also show on the difference dial as fractions of a cycle.

10.1.2   Difference Clock Changes Direction of Rotation When Doppler Changes Sign

When the SV passes directly overhead, the Doppler rate goes through zero and changes sign. The observer of the difference dial would see it slow down to zero-rate and then reverse its direction of rotation as the Doppler changes sign.

In this model, we can make a somewhat arbitrary choice about Doppler signs and direction of rotation of the Difference dial. Here we choose CW rotation for an approaching SV, which would have a positive Doppler. For a receding SV, negative Doppler, the rotation direction is then CCW.

10.1.3   Full Cycle Counting, Partial Cycles, and Sign Issues

The Difference clock dial can only record one cycle of range or movement in range with each cycle equal to one wavelength of $L1$ or 0.190293672798 m. In order to measure larger ranges we need to count full cycles that have passed on the Difference dial. The Cycle counter does this for every time the Difference Dial crosses its zero timing mark.

At each zero mark crossing, the sign of cycle to be added must also be determined. For a positive Doppler, we need to subtract a cycle as Range is closing (shrinking). For a negative Doppler, we need to add a cycle as range is opening (growing).

Partial cycles are measured as indicated in Figure 10.1.

The issue of resolving the sign of the difference clock rotation can be avoided by adding a *bias* rate to the difference clock. This is, of course, how most receivers do this operation. The bias rate is simply the last (nominal) IF frequency in the conversion chain. With a bias rate, the difference dial only rotates in one direction with the Doppler present as a small *phase&rate* variation about the bias rate.

Another way to think about it is to realize that in the method presented here, our difference dial is a *BaseBand* dial, i.e., no carrier is present. When the IF method is used, the Carrier Phase Dial's *phase&rate* information is *carried* as a small delta modulation on the final IF rate.

10.1.4   Range Measurement Using Full and Fraction Cycles of Difference Dial, Integrated Doppler

At any instant of time, the full cycle counter and the instantaneous phase of difference dial with respect to its zero timing mark can be combined to make a measure of receiver to SV range accurate to a fraction of a wavelength of the *L*1 carrier. The sum of the full cycle counter and the fraction of cycle indicated on the dial are multiplied by 0.190293672798 m to obtain the range measurement at any instant.

By definition, the difference dial is performing integration on the Doppler Rate that appears on it. With the ability to count full cycles of the difference Dial, a pure form of *Integrated Doppler* (which is a range measurement) is produced, one that is free of errors or Bias rates as we assumed our local L1 reference was perfect.

In practical receivers a *pure* integrated Doppler is never produced due to bias IF rates and errors of rate on the LO's used for down conversion which are just multiplied up Master Clock rate errors. Thus when practical receivers implement "Doppler integration" or more informatively *Total Integrated Doppler* (a.k.a. *Total Integrated Carrier Phase*), they typically have bias and errors present in the result. These biases and errors must be subtracted away in order to arrive at an estimate of just the Integrated Doppler term.

10.1.5   The Initial Value of the Full Cycle Counter

When the receiver starts tracking an SV, it does *not* know the delta time or range to the SV accurate to a fraction of a L1 carrier cycle or wave length. For an L1 only receiver, this range for a rising SV is most likely accurate to a few tens of meters as this measurement is based on Code Phase measurements. This uncertainty in range translates to possibly hundreds of cycles of L1 carrier. The initial value that is loaded into the full cycle counter at the start of Carrier Phase measurements could be the range estimate as provided by Code Phase measurements, it would typically be the closest integer number of *L*1 wavelengths of that range.

Single frequency receivers have little hope of resolving the uncertainties or *ambiguities* of the initial full cycle count for Carrier Phase-based measurements. Differential use of carrier phase measurements in *L*1 receivers is more likely as the difference operation can eliminate the initial full cycle issue by (hopefully) subtracting it away.

10.1.6 *Practical Issues*

L1 receivers *do not* typically use a direct comparison of a local L1 clock (or reference) to the *received* L1 signal for Carrier Phase Dial range measurements. Rather the received L1 carrier is down-converted to an IF frequency where a replica dial is phase locked to this IF frequency. The replica clock dial produced by this phase lock process is essentially a difference dial (when locked), just as presented here, but with a bias rotation rate added equal to the nominal IF rate. In addition, the phase reference used will be based of the master oscillator as discussed in Chap. 8, and not a clock at the *L*1 rate as used in the mechanical model presented here.

As seen in the Zarlink receiver in Chap. 9, a direct full cycle counter was attached to the NCO output. The NCO itself was in phase lock with the down-converted IF from the received L1 signal. At the TIC instant, the full and partial cycle information was captured by the Cycle Counter and NCO phase, respectively. This method and its software-based variants is typically how the carrier phase is implemented in L1 receivers.

Lastly it should be clear that the direction of the difference dial is used in determining the sign of the Doppler. In the receiver processing, this nearly always means some form or quadrature or *I* and *Q* signal processing is needed to extract not just the magnitude of the Doppler shift but the sign of it as well. This statement applies to phase angle as well.

10.1.7 *Cycle Slips*

As practical receivers do not use the L1 Carrier Phase Dial directly as received, but rather create a replica dial by a phase lock process, there is always the potential problem of lock being lost and the phase cycle count being corrupted by cycle slips.

When such slips occur, they will typically be an integer number of ½ cycles for L1 carrier tracking. This occurs because the L1 carrier being tracked has 50 Hz bi-phase data left on it. Depending on where the carrier loop picks up lock again (i.e., which 50 Hz data bit), there can be a 180° or ½ cycle change from the last steady lock state.

10.2 L1 Carrier Loop Processing

A model of typical L1 carrier loop processing is shown in Fig. 10.2. This model can apply to any of the carriers used in the GPS transmission. The carrier information leaves the SV with no noise and zero Doppler. By the time it reaches the receiver, the carrier signal is changed in its perceived frequency (Doppler), has delay due to

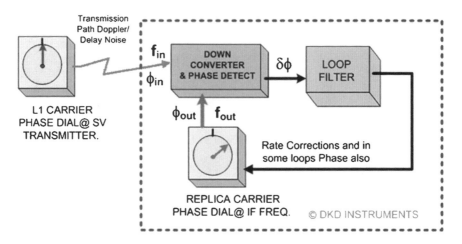

Fig. 10.2 Typical carrier loop processing

path length, and significant noise due to very low received signal power. In this subsection, we will assume no noise and stable (rate and phase) signals both received and self-generated (i.e., LO's) in the receiver.

It is the carrier loop's job to provide a *replica* carrier phase dial that has minimal errors of phase and frequency with respect to the incoming carrier. The carrier recovery process is accomplished by using a phase-locked loop. The loop will suppress noise and when locked provide a recovered (replica) Carrier Phase dial that we can use to make range measurements.

In our model, we have combined two functions in one block that are typically shown as two blocks. Most GPS receivers use an IF frequency to process the carrier signal. Here we have combined the down-conversion and the phase error detector in one block.

10.2.1   Analog Base Band Model of L1 Carrier Loop

The IF loop from above is now expressed as a control loop operating on the phase variable, ϕ, see Fig. 10.3. The input carrier information is contained in ϕ in and when the loop is locked the difference between ϕ in and ϕ out, $\delta\phi$, should be small. The loop filter smoothes the phase error, $\delta\phi$, for use on controlling the rate (and phase) of the carrier replica dial (oscillator).

In the base band model, a switch has been added to allow the loop to "open". The switch is set to the track position until time t_0 and then it is set to zero. If we assume that the loop was locked and stable before the switch was moved at time t_0, what will happen to the loop?

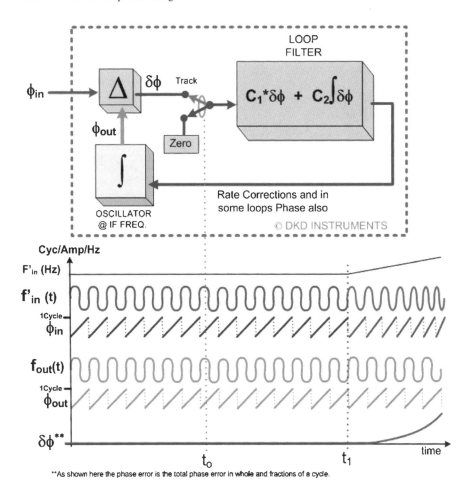

Fig. 10.3 Baseband model of L1 carrier tracking loop

10.2.2 Loop Opened, Input Signal (Fin) is Constant, Loop Filter Input Is Zero

At t_0, the loop is opened and the input to the loop filter is set to Zero. As the loop was assumed locked, the integrator in loop filter will retain the proper setting for the rate (and possibly phase) of the replica clock dial or oscillator. As long as the input signal does not change in rate or phase and there is no noise in the system, $\delta\phi$ will remain zero (all rates and phases perfect). The time signals of Fig. 10.3 reflect this and no change is seen in any signal.

10.2.3   Loop Opened, Frequency Ramp on Input Signal at t_1

At time t_1, frequency of the input signal starts to change. In particular, it starts a steady increase with time. This is, of course, what would be observed for a received GPS SV signal, as its Doppler does not remain fixed but changes with time, with largest Doppler magnitude as SV rises and then passing through zero as SV passes overhead then reversing sign and increasing as it descends.

With the switch still set to zero at the loop filter input, the loop cannot react. We now see the phase error, $\delta\phi$, start to grow with time in response to the changing phase of the input signal against the (static) phase of its local clock. As seen in Fig. 10.3, the change in input frequency after t_1 is exaggerated for clarity and the phase, $\delta\phi$, is not scaled.

10.2.4 $\delta\phi$ Is an Estimate of SV Acceleration with Respect to Receiver (on LOS)

Before the input signal changed, $\delta\phi$ was zero in our perfect receiver model. As soon as there was a change in the input signal, ϕ_{in}, $\delta\phi$ then reacted. We conclude that that an observed change in $\delta\phi$ (when in lock and steady state) can only occur if the observed Doppler has a non-zero first derivative. In other words, in our steady state, noiseless model, $\delta\phi$ becomes non-zero when there is an acceleration of the SV with respect to the receiver.

10.2.5   Loop Filter Is a Weighted Sum of Acceleration and Velocity Terms

The loop filter contains the integral of $\delta\phi$, which is an estimate of phase rate or $\phi'(t)$. It is also an estimate, within a bias offset typically, of SV velocity (or Doppler) with respect to receiver (LOS). The integral of $\delta\phi$ may contain a constant bias (or IF frequency) that must be subtracted out to obtain just the Doppler term due to relative movement between SV and receiver. The weighted linear combination of the acceleration (Doppler rate) term and the velocity term (Doppler) are used as control inputs to the oscillator, which performs an integration operation producing the output phase.

With this in mind, we can see that the output phase ϕ_{out}, which is proportional within a constant offset to SV range, will not change if the acceleration (proportional to $\delta\phi$), and the velocity(proportional within a constant offset to $\int\delta\phi$) is zero. This is as expected and tells us that SV to receiver range at that instant would be static.

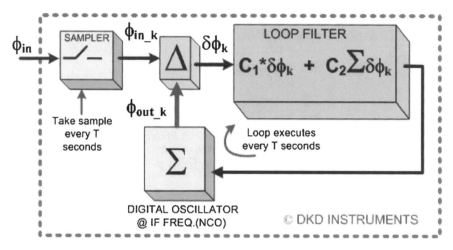

Fig. 10.4 All digital carrier loop or carrier tracker

10.2.6 All Digital Baseband Carrier Loop

In Fig. 10.4, the analog base band loop of Fig. 10.3 is replaced by a sampled time version. The analog integrators have been replaced by summations. The loop executes once every T seconds, where T is typically in the 1–20-ms range. The subscript, k, refers to the loop sample number which is incremented every T seconds.

In steady state operation the phase error, $\delta\phi_k$ would not change if the input, ϕ_{in} was static, noise free, and an exact integer multiple of the discreet phases possible in the Digital Oscillator. If over the time interval T the input phase changes, the next sample of $\delta\phi_k$ will reflect this change. As before, we see that $\delta\phi_k$ is an estimate of SV acceleration with respect to receiver (LOS).

10.2.7 Typical Loop Up-Date Relationships

The current interval is denoted by k, the next value by $k + 1$. The summation of $\delta\phi_k$ in (10.2) is over all past and present values of $\delta\phi_k$ from the beginning of the SV track, $k = 0$, to the present loop sample time, k;

$$\delta\phi_k = \phi_{in\_k} - \phi_{out\_k.../} \times \text{current phase error} \qquad (10.1)$$

$$\phi_{out\_k+1} = \phi_{out\_k} + C_1 \times \delta\phi_k + C_2 \times \Sigma\delta\phi_{k.../} \times \text{next value of phase output} \qquad (10.2)$$

From (10.2), we see that the new value of the output phase is equal to the last value of output phase plus a weighted sum of past and current values of $\delta\phi$. As noted above, $\delta\phi$ is ~ proportional to acceleration of the SV along LOS to the receiver.

The term $\Sigma\delta\phi$, a sum of all $\delta\phi_k$ from $k = 0$ to the present value of k, is roughly proportional, within a constant or bias, to the velocity of the SV along the LOS to the receiver. If we let $x(nT)$ be the SV to receiver range at time nT, we can see that (10.3) is really an estimate of the following equation;

$$x([n + 1]T) = x(nT) + C_2 \times x'(nT) + C_1 \times x''(nT) \tag{10.3}$$

Where $x([n + 1]T)$ is the next range, $x(nT)$ is the current range, $x'(nT)$ is the current velocity (LOS) and $x''(nT)$ is the current acceleration (LOS). The units of C_1 and C_2 would have to be adjusted to produce a distance after the multiply operation.

10.2.8   Initial Value of $\Sigma\delta\phi_0$

When the loop is started up, the first values must be initialized. Setting them to zero can be done but typically one initial value is not set to zero and that is the very first value of $\Sigma\delta\phi$, or $\delta\phi_i$. $\delta\phi_i$ is often set equal to the number of cycles of F'_{in} that would occur in a Loop update period T;

$\delta\phi_i = F'_{in} \times T$ where F'in is the nominal, zero Doppler rate of input sig. (IF freq.)

This is done so that the sum of $\delta\phi_k$, an approximation of next the phase rate, is now properly biased to be directly used as a frequency command to Digital Oscillator (NCO) every T seconds.

10.2.9   Slope of Accumulated Carrier Phase Can Be Reversed (i.e., Inverted Plot)

Depending on how the Carrier Phase is accumulated, the resulting plot of the phase data may show an inversion with respect to true range data. This would occur in the Zarlink method as the phase accumulated is directly tapped off the NCO output. The output rate of the NCO will be higher for an approaching SV due to a positive Doppler. Therefore the accumulated phase total will be *increasing* when the actual range is *decreasing*. In such a case, if we offset and scaled the Carrier Phase data as range and compare against a true range plot we would see that the Carrier Phase range plot is inverted.

10.2.10   Accumulated Phase Is Typically the Total Phase

When the phase is accumulated from an locked dial, it is usually accumulated as the number of full cycles of the IF dial or clock that have occurred since start of the track. This sum is normally comprised of full cycles and the fraction of cycle at the measurement instant. Because the sum is counting Nominal IF plus Doppler Offset plus first LO Error, the total phase at any instant has offsets and errors. In addition, the higher the Nominal IF, the larger the number of cycles will be accumulated over the track time.

10.2.11 To Extract Measured Phase Subtract Number of Cycles of the Nominal IF Rate

If the accumulated phase has a IF bias rate in it and we wish to just have the portion due to Doppler and LO errors we need to subtract off the total number of nominal IF cycles that would have been added to the total at the time instant of interest. If that total time is ΔT then the number of cycles we need to subtract off is $\Delta T \times F'$in, where F'in is the nominal IF rate as above. At this point, the accumulated phase still has LO rate errors in it. Those must be accounted for in computations for the rate error on the master oscillator. Once the LO rate error over time is known, the accumulated phase could be further corrected.

10.2.12 Units of Accumulated Carrier Phase Can Be in Cycles, Counts, Etc

Typically the accumulated phase is expressed in cycles but not always. Sometimes it is scaled in NCO counts and other units. For example, if the NCO used has a 32-bit phase register then one cycle of phase is equal to 2×32 counts. The accumulated phase could be expressed in counts as scaled to the NCO being used in the loop. Make sure you understand the scaling if you use the phase data!

10.3 Using the L1 and L2 Carrier Phase Dials to Create a New Dial

In Receivers that capture L1 and L2 transmissions from the SVs, it is possible to create a new dial for use in range (phase) measurements. Figure 10.5 shows a model of the L1 and L2 carrier dials at the SV before being transmitted. The two dials are synchronous in that they are derived from a common reference frequency of 10.23 MHz. Because of this synchronous relation, we create a new dial in our receiver.

To create the new dial we can use the mechanical models developed in Book I. In particular, if we connect the L1 and L2 carrier phase dials to a differential

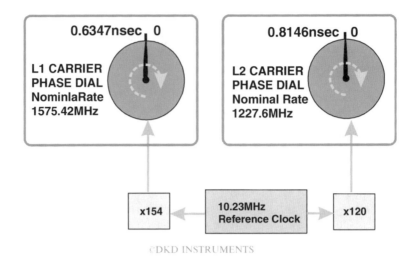

Fig. 10.5   L1 and L2 carrier generation at the SV

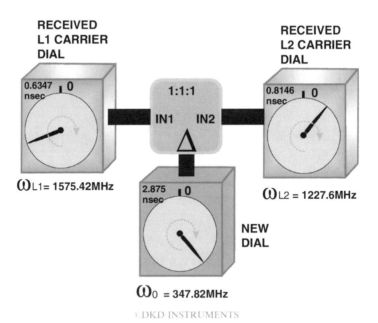

Fig. 10.6   A mechanical model of L1/L2 receiver new dial creation from L1 and L2 dials

gear block as shown in Fig. 10.6, we will create a new dial that has a rate of rotation that is equal to the difference in rates of L1 and L2 dials. All the rates shown are nominal receiver rates referenced to GPS rate and have zero Doppler on them.

The new dial has a nominal rate of 347.82 MHz. The new dial will have phase noise from the receiver processes present on the L1 and L2 dials. Assuming the 1/100 cycle precision we see that the new dial can resolve down to about ½ inch or 12 mm. The un-ambiguous range of the new dial is ~2.87 feet. This new dial can also help resolve integer cycle ambiguities present in L1 and L2 carrier phase measurements.

To create this dial in receiver signal processing, we could multiply the NCO output from the L1 carrier loop against the NCO output from the L2 carrier loop. The output of the multiplier would be low-passed (or accumulated for DSP) to remove the sum component.

10.4   Analysis of the Measurement and Use of Total Integrated Carrier Phase $\Phi(t)$

The goal of this section is explain a number of questions regarding choices made in the Turbo Rogue signal processing and to explore what is called Total Integrated Carrier Phase. In particular, answers to the following questions are sought;

- In a dual frequency receiver what values should be used for the final IF frequency of $L1$ and $L2$?
- Why does the Turbo Rogue receiver use the ratio 154/120 in the final IFs?
- What happens when we subtract the Integrated Phase of $L1$ and $L2$?
- What happens when take the ratio of the Integrated Phase of $L1$ and $L2$?

Along the way in the effort to answer the above questions we will develop the needed relationships and equations to formulate and manipulate the measured phase values from $L1$ and $L2$ processing. This includes the single carrier phase differences, between carrier phase differences and ratio of $L1$ and $L2$ carrier phases.

To be able to answer the above questions requires a fair amount of mathematics and many variables. The first order of business is to detail the variables used in the following discussion.

Variable Definitions

$v(t)$ = LOS velocity
$x(t)$ = Accumulated range by integrating $v(t)$
x_0 = Range offset at start of integration interval
$L1$ = 1,575.42 MHz, L1 carrier Frequency
$L2$ = 1,227.6 MHz, L2 Carrier Frequency
$L3$ = $L1$–$L2$ = 347.82 MHz, Synthetic L3 Carrier Frequency
$F_{d1}(t)$ = Doppler frequency offset on L1 Carrier
$F_{d2}(t)$ = Doppler frequency offset on L2 carrier
F_{c1} = Final Carrier/IF frequency after all conversions for L1 processing
F_{c2} = Final Carrier/IF frequency after all conversions for L2 processing
α/β = $L1$ to $L2$ frequency ratio, $L1/L2$ (154/120 for GPS)

σ/δ $\quad = F_{c1}$ to F_{c2} IF Frequency ratio, F_{c1}/F_{c2} (any(?) ratio where $\sigma > \delta$)

$\Psi_{d1}(t) =$ Integrated Doppler for L1, i.e., Doppler phase

$\Psi_{d2}(t) =$ Integrated Doppler for L2, i.e., Doppler phase

Θ_{c1} $\quad =$ Integrated carrier phase from F_{c1}, units are cycles of F_{c1}

Θ_{c2} $\quad =$ Integrated carrier phase from F_{c2}, units are cycles of F_{c2}

$\Phi_1(t)$ $\quad =$ Total integrated phase for L1 processing

$\Phi_2(t)$ $\quad =$ Total integrated phase for L2 processing

$\Phi_3(t)$ $\quad =$ Total integrated phase for synthetic L3 processing

$e1$ $\qquad =$ Phase offset at start of integration for total $L1$ phase

$e2$ $\qquad =$ Phase offset at start of integration for total L2 phase

K_1 $\qquad = 1{,}575.42$ MHz/Speed of light (i.e., Doppler scale const for L1 carrier)

K_2 $\qquad = 1{,}227.6$ MHz/Speed of light (i.e., Doppler scale const for L2 carrier)

m $\qquad =$ Total number of wave lengths of $L1$ contained in range

n $\qquad =$ Total number of wave lengths of L2 contained in range

l $\qquad\;\, =$ Total number of wave lengths of L3 contained in range

M $\qquad =$ Integer or whole number portion of measured $\Phi_1(t)$

N $\qquad =$ Integer or whole number portion of measured $\Phi_2(t)$

Note that the following ratios are equal to α/β (i.e., 154/120):$F_{d1}(t)/F_{d2}(t), K_1/K_2, m/n$

10.4.1 Analysis of Total Integrated Carrier Phase and Its Uses

It will be assumed for this analysis that all frequencies are exact, specifically that F_{c1} and F_{c2} have no frequency errors due to small offsets of local reference oscillator, i.e., the local reference oscillator that is used to generate all LOs used to down convert to final IF frequencies is assumed perfect.

The quantities $\Phi_1(t)$ and $\Phi_2(t)$, from (10.2) and (10.3) below, represent what is called in this text the Total Integrated Carrier Phase. Typically, in a receiver, only the values of $\Phi_1(t)$ and $\Phi_2(t)$ are what is directly measured. In other words, the components that make up this total phase can only be inferred by computations on these measurements. This issue of non-separable phase components occurs from the fact that at the final IF for L1 and L2 signal processing, the Doppler Offset and the final IF carrier frequency appears as a sum, see Fig. 10.7. It is this sum that is integrated. This section attempts to explain some of those computations and the estimates they produce of range, range rate , use for cycle slip detection , ambiguity resolution, and most importantly the effect of choosing the final IF ratio σ/δ. For a model of how the Total Integrated Carrier Phase measurements can be done, see below.

It seems clear that the choice to have the final IF frequency ratio the same as the ratio of $L1$ to $L2$ seems only to effect the *ratios* of the Total Integrated Phase quantities, namely $\Phi_1(t)/ \Phi_2(t)$, $\Phi_1(t)/ \Phi_3(t)$ and $\Phi_2(t)/\Phi_3(t)$. In particular, the difference relations remove the integrated carrier phase component contribution from the nominal IF frequency. In other words, the difference relations seem to get rid of any effects of a final IF ratio not equal to 154/120.

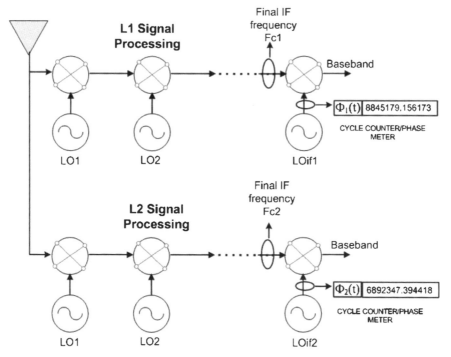

The Total Integrated Carrier Phase Measurements of $\Phi_1(t)$ and $\Phi_2(t)$ can be done at the last LO, shown above as LOif1 and LOif2. This last LO is phase locked to incoming signal(via CarrierTracking Loop) such that the output of the last mixer is at Baseband. In other words the last LO is equal to the last IF , aka Fc1 & Fc2

If Fc1/Fc2 = 154/120 then $\Phi_1(t)/\Phi_2(t)$ = 154/120 (as shown above)

Fig. 10.7 Generic L1/L2 processing

The analysis below also indicates that the ratio $\Phi_1(t)/\Phi_2(t)$ is only equal to a constant when the ratio of the final IF carrier frequencies used for L1 and L2 processing (F_{c1}/F_{c2}) is equal to the ratio of $L1$ to $L2$ frequencies, in other words $\sigma/\delta = \alpha/\beta = 154/120$. By having $\Phi_1(t)/\Phi_2(t) = \alpha/\beta$, a fixed constant, allows simple comparison of measured $\Phi_1(t)$ or $\Phi_2(t)$ values by either prediction of one from the other or just verifying that the ratio of the two measurements is correct by examining the ratio. It seems likely that such an easy cross check of the measured values of $\Phi_1(t)$ and $\Phi_2(t)$ would be very useful in resolving the carrier wavelength ambiguity or in the detection of cycle slips.

This analysis does not pursue the details of resolving the integer wavelength (or cycle) counts contained in the range or the impact of σ/δ not being equal to 154/120 on the ratios involving $\Phi_3(t)$.

10.4.2   Derivation of Total Integrated Carrier Phase

At the final IF processing for L1 and L2 signal processing, we have the two carriers with frequencies;

$$[F_{c1}+F_{d1}(t)](\text{for L1 processing}) \text{ and } [F_{c2} + F_{d2}(t)] \text{ (for L2 processing)} \quad (10.4)$$

If we integrate the expressions in (10.4), we get:

$$\Phi_1(t) = \Theta_{c1} + \Psi_{d1}(t) + e_1 \text{for L1 processing} \quad (10.5)$$

$$\Phi_2(t) = \Theta_{c2} + \Psi_{d2}(t) + e_2 \text{ for L2 processing} \quad (10.6)$$

If we set the starting phases to zero for both L1 and L2 processing and start the integration of L1 and L2 at the same instant:

$$\Phi_1(t) = \Theta_{c1} + \Psi_{d1}(t) \text{ for L1 processing} \quad (10.7)$$

$$\Phi_2(t) = \Theta_{c2} + \Psi_{d2}(t) \text{ for L2 processing} \quad (10.8)$$

and for this same condition (starting phases set to zero/same start instant);

$$\Theta_{c1}/\Theta_{c2} = \sigma/\delta \text{ and } \Psi_{d1}(t)/\Psi_{d2}(t) = \alpha/\beta \quad (10.9)$$

10.4.3   Differences and Ratios of $\Phi(t)$'s

Now that we have expressions for the total integrated carrier phase for L1 and L2 processing, we can explore some of the ways they can be combined to calculate different measurements/estimates from them. The most common manipulations are single differences and ratios.

10.4.4   The Ratio of $\Phi_1(t)$ to $\Phi_2(t)$

Forming this ratio seems to be the most important clue in understanding why the choice of keeping the final IF ratios in the Turbo Rouge receiver equal to 154/120 was made. Using (10.7) and (10.8) we form the ratio;

$$\Phi_1(t)/\Phi_2(t) = [\Theta_{c1} + \Psi_{d1}(t)]/[\Theta_{c2} + \Psi_{d2}(t)] \quad (10.10)$$

by using relationships shown in (10.9) we can form;

$$\Theta_{c2} = [\delta/\sigma] \times \Theta_{c1} \text{ and } \Psi_{d2}(t) = [\beta/\alpha] \times \Psi_{d1}(t) \tag{10.11}$$

Substituting these relations into (10.10);

$$\Phi_1(t)/\Phi_2(t) = [Q_{c1} + \Psi_{d1}(t)]/[(\delta/\sigma) \times \Theta_{c1} + (\beta/\alpha) \times \Psi_{d1}(t)] \tag{10.12}$$

if $\sigma/\delta = \alpha/\beta$ then (10.12) reduces to;

$$\Phi_1(t)/\Phi_2(t) = \alpha/\beta \tag{10.13}$$

Equation (10.13) shows that for the case where $\sigma/\delta = \alpha/\beta$, the ratio of $\Phi_1(t)/\Phi_2(t)$ is equal to a simple known constant, i.e., the ratio 154/120.

By having this simple relationship for the ratio of $\Phi_1(t)/\Phi_2(t)$ would seem to be very useful for many verification and comparisons as already mentioned above. It should be stressed that for this relation to hold, the integrations resulting in the $\Phi_1(t)$ and $\Phi_2(t)$ measurements must be started at the same time from a zero phase offset with respect to each other. In addition, if a cycle slip occurs in either L1 or L2 carrier tracking, the ratio should detect this and corrections must be made (or restart the integration) for the ratio to continue to hold up.

A simulated, growing parabolic Doppler offset was done using Xcel and the ratio $\Phi_1(t)/\Phi_2(t)$ was computed for two cases: $F_{c1} = 308$ kHz, $F_{c2} = 240$ kHz and $F_{c1} = 308$ kHz, $F_{c2} = 275$ kHz. As expected in the first case, the phase ratio was a constant at 1.2833 and in the second case it was a parabola plot that tracked the parabolic Doppler offset, see Plot 10.1.

10.4.5 Single Carrier Difference of $\Phi(t)$, or $\Delta\Phi(t)$

In this case, measurements from L1 or L2 processing are differenced against themselves. Typically, the proceeding measurement is subtracted from the current measurement. For the L1 carrier we could write;

$$\Delta\Phi_1(t_n) = \Phi_1(t_n) - \Phi_1(t_{n-1}) \tag{10.14}$$

using (10.2) we see that;

$$\Delta\Phi_1(t_n) = \Psi_{d1}(t_n) - \Psi_{d1}(t_{n-1}) \tag{10.15}$$

(The constant phase terms cancel)

The constant phase terms are subtracted away and we are left with a difference of two integrated Doppler terms. We can express this directly in range by noting that;

$$\Psi_{d1}(t) = \int [K_1 \times v(t)] = K_1 \times x(t) + x_0 \tag{10.16}$$

Ratio of Phase of L1 path to Phase in L2 Path

Plot 10.1 Ratio of $\Phi_1(t)/\Phi_2(t)$ for IF frequency ratio of 1.12

Using this in (10.12);

$$\Delta\Phi_1(t_n) = K_1[x(t_n) - x(t_{n-1})] \tag{10.17}$$

(The constant range terms cancel)

We can now see that if we difference two Integrated Phase measurements from two different times, t_n and t_{n-1}, from the same carrier, we get an estimate of the change in range that occurred in that time interval. This measurement of the change in range is extremely accurate as it is based on carrier phase measurements instead of code phase measurements.

By doing this same computation for the L2 carrier for the same time instants, we can produce another estimate of the same change in range.

10.4.6 *Between Carrier Phase Differences, $\Phi_1(t): \Phi_2(t)$ (ratios and single differences)*

If we difference the total integrated phase measurements for the two carriers, L1 and L2 at the same time instant, we are essentially creating a new carrier; call it L3. Using (10.5) and (10.6), we take the difference between them and get;

$$\Phi_3(t) = \Psi_{d1}(t) - \Psi_{d2}(t) + \Theta_{c1} - \Theta_{c2} \qquad (10.18)$$

We got rid of e_1 and e_2 by setting the phase to zero for both $\Phi_1(t)$ and $\Phi_2(t)$ at the start of the integration interval. This new synthetic carrier would have a frequency of 347.82 MHz, the difference frequency between L1 and L2, if it were actually transmitted (see Sect. 10.3 above).

One use is to help in determining the total number of carrier wavelengths of L1 and L2 in a given range, i.e., m and n. With its lower frequency, there are far fewer wavelengths of L3 in a given range compared to L1 and L2. This information helps shrink the number of possible values of m and n values to search through. The single letter L denotes the total number of wavelengths of L3 in the range.

The ratio of L1 and L2 frequencies to L3 frequency is;

$$\Gamma_{13} = L1/L3 = 4.52911 \text{ and } \Gamma_{23} = L2/L3 = 3.529411$$

Using the relationship developed from (10.9) we see that;

$$\Phi_3(t) = x(t)[K_1 - K_2] - [\Theta_{c1} - \Theta_{c2}] \qquad (10.19)$$

If we differenced (10.19) at two different measurement times, t_n and t_{n-1}, we get rid of the constant phase term $[\Theta_{c1} - \Theta_{c2}]$, just as we did in (10.15) above, and form the single difference phase calculations from our synthetic carrier L3;

$$\Delta\Phi_3(t) = [K_1 - K_2] \times [x(t_n) - x(t_{n-1})] \qquad (10.20)$$

Now we have a third estimate of the change in range for the time interval from t_{n-1} to t_n.

10.5 Time Precision and Time Resolution in the Turbo Rogue Receiver

This section investigates the Time Resolution and Time precision of Turbo Rogue receiver. Estimates of internal as well applied signal phase noise and jitter are presented. A fairly detailed analysis of the effects of phase error sampling jitter, or sample-induced phase noise, is presented. A brief estimate is presented for the incoming phase jitter on received signal due to finite SNRs.

Table 10.1 Conversion of timing Jitter in pico seconds to Distance in mm

σ_t Stand. Dev. in Pico Sc	σ_d Stand. Dev. Equivalent of σ_t expressed as distance in Millimeters
100.0000	30.0000
50.0000	15.0000
30.0000	9.0000
20.0000	6.0000
10.0000	3.0000
5.0000	1.5000
3.0000	0.9000
1.0000	0.3000
0.5000	0.1500

Conversion formula is $\sigma_d = \sigma_t \times$ Speed of Light

When considering time measurements in the Turbo Rogue receiver, it is important to keep in mind that the receiver makes phase measurements against its own internal clock that when scaled by speed of light result in range measurements. In other words, phase difference measurements are translated to time measurements, which then are translated to range measurements. To come full circle, we could have asked what is the ultimate range resolution and precision of the Turbo Rogue receiver? The result is the same in that we must address the ultimate phase/time resolution and precision of the receiver. In addition, we must keep in mind that time resolution does not necessarily directly translate to time precision. Noise and other effects are often at higher levels than the measurement resolution. Thus our measurement precision is usually less than our measurement resolution. This is of course true for many measurement processes.

Turbo Rogue typically makes two basic types of phase measurements:

Code Phase (C/A and P code)
Carrier Phase (L1 and L2 carriers)

Of these two types of measurements, Carrier phase typically has 1–3 orders of magnitude more time resolution and precision than code phase measurements. This follows directly from the fact that the P code rate is ~1/100th the rate of L1 and L2 carriers while C/A code rate is ~1/1,000th the carrier rate. The increased precision in time interval measurement is directly proportional to the frequency of the signal being measured. For carrier phase measurements, we are talking about time precisions measured in the pico second range (10×-12)! See Table. 10.1 for typical picosecond time jitters (as standard deviations) involved to equivalent range uncertainty.

With this in mind we can reformulate the time resolution/precision question slightly differently. In particular even though code phase measurements are very important to the receiver, the receiver's ultimate performance in terms of phase (or equivalently time) resolution is determined by a carrier phase type measurements. If we know the ultimate carrier phase resolution/precision, say in fractions of a cycle, then we simply multiply by the time of one cycle of the carrier ($L1$ at 1,575.42 MHz or $L2$ at 1,227.6 MHz) too get the equivalent time resolution/precision.

The carrier phase disturbance effects discussed here are modeled as noise processes. To pursue a measure of carrier phase precision, we need a convenient measure of these noise processes. Specifically, we will use the mean square values of the noise processes that disturb the carrier phase. Further, we will assume that these noise processes possess a zero mean, thus the square root of the mean square value is the familiar standard deviation of the noise (or time/phase jitter) under discussion.

10.5.1 Disturbances to Carrier Phase, An Overview

Now that we know it is carrier phase that enables the ultra high-resolution measurements in the Turbo Rogue receiver, we need to form the larger picture of just what can corrupt carrier-received carrier phase. By understanding the sources of phase disturbance, we can start to estimate the limits these disturbances place on carrier phase measurement precision.

First we can imagine a basic partitioning of phase disturbances into receiver system phase disturbances and disturbances external to the receiver. The external disturbances are typically associated with the atmosphere phenomenon between the receiver and the satellite. To quantify such disturbances, we will describe them as time jitter, σ_t, where σ^2_t is the mean square value of the noise process producing the observed time jitter (for zero mean σ_t = standard deviation).

If we express the jitter sources that are part of receiver system receiver as σ_{t_int} and jitter sources external to the receiver as σ_{t_ext}, we can express the total mean square jitter as:

$$\sigma^2_{t total} = \sigma^2_{t\_int} + \sigma^2_{t\_ext} \tag{10.21}$$

Further, we can break down the term $\sigma_{t\_int}{}^2$ into its constituent parts (a partial list):

$$\sigma^2_{t int} = \sigma^2_{t\_LO\_} + \sigma^2_{tq} + \sigma^2_{t\_AntCable} + \sigma^2_{t\_Temp\_} + \sigma^2_{t\_RF\_} + \ldots\ldots + \sigma^2_{t\text{-misc}} \tag{10.22}$$

Where,

$\sigma^2_{t\_LO\_}$ = Mean square value of Jitter related Local Oscillator Phase Noise

$\sigma_{tq}{}^2$ = Mean square value of Jitter due to sampling error of incoming carrier zero crossing instant

$\sigma^2_{t\_AntCable}$ = Mean square value of phase disturbances caused by antenna preamp, cable from antenna to receiver, etc.

$\sigma^2_{t\_Temp}$ = temperature effects on cables, etc.

$\sigma^2_{t\_RF\_}$ = variation in phase in RF down converter with temperature, mechanical vibration

$\sigma^2_{t\text{-misc}}$ = remaining contributors not listed

Of all the jitter terms listed above, it can be shown that σ_{tq} is one of the largest, if not the largest, contributor to the internally generated timing jitter that degrades the carrier phase measurements, see Fig. 10.8. Estimating the effects that σ_{tq}^2 will have on the precision of Turbo Rogue carrier, phase measurements will dominate the remainder of this section.

10.5.2   Turbo Rogue Time Resolution

Referring to Fig. 10.8, we see that the carrier-tracking loop uses an NCO to match the incoming phase and frequency as seen by the receiver. The exact frequency produced depends on Doppler and the nominal IF frequency (308 kHz $L1$, 240 kHz $L2$). The number of bits in the phase register of the NCO sets the phase resolution and hence carrier phase (time) resolution. The Turbo Rogue processor uses a 24-bit NCO for both L1 and L2 Carrier tracking. This works out to an approximate time resolution for L1 processing of 0.2 ps (Time Res ~ (1 Cycle/2 × 24)/308 kHz). The resolution will be slightly lower for L2 processing, as L2 IF is 240 kHz nominal. Strictly speaking, the carrier tracking NCO time resolution is only achieved after significant averaging or observation time due to the quantization error of the NCO, which is clocked at 20.456 MHz (NCO Phase is quantized to ~48 ns).

10.5.3   Turbo Rogue Time Precision

Estimating the time precision is a much more difficult task than estimating the time resolution. To help understand the issues involved, consider the following experiment. We disable the C/A or P code part of the Turbo Rogue receiver thereby configuring it to receive an un-coded sine wave receiver (i.e., a CW receiver). We now feed in a *perfect* sine wave into the receiver, one that has zero phase and rate errors with respect to the receiver's local clock (and very low phase noise). We then take carrier phase data and examine it. Would the precision in phase data (when converted to time data) be near 0.2 ps as predicted from the time resolution calculation above?

Typically no, we will not see 0.2 ps of time precision in reported or observed carrier phase data, at least not without a lot of processing/averaging of the reported carrier phase data. Rather we would see carrier phase data, if viewed at high resolution, which can have noise with a magnitude above the 0.2 ps level. In others words, our ultimate time resolution can be masked by *internally* generated phase disturbances. To sum up this observation, we can state: *The magnitude of*

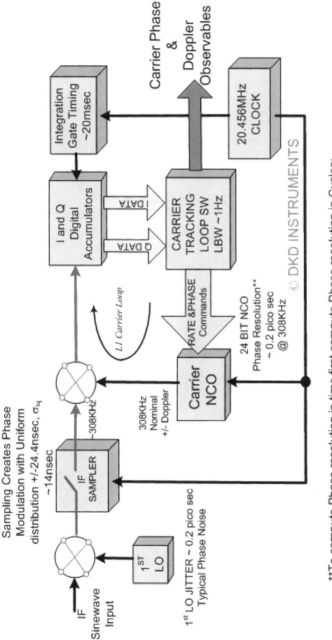

Fig. 10.8 Analysis of Turbo Rogue timing resolution/precision with sine-wave input

the reported carrier phase measurement noise, with a perfect input, determines the ultimate time precision of the Turbo Rogue receiver.

In other words, to answer the time precision question, we must know the *carrier phase data noise floor*. To estimate the carrier phase data noise floor, we must estimate the Turbo Rogue's own internal sources of phase disturbances. In identifying the Turbo Rogue's internal phase disturbance sources (that would corrupt an incoming perfect carrier's phase), we can estimate the limits of time precision.

10.5.4   Internally Generated Phase Noise in the Turbo Rogue Receiver

Figure 10.8 shows a simplified model of the Turbo Rogue carrier processing, including the carrier tracking loop. The received input is down-converted to an IF by the first mixer. The subsequent mixer down-converts the signal to base-band and is done using hardware/digital methods. Before the Digital down-convert, the analog IF is sampled at 20.456 MHz. The base-band output signal from the last mixer is used in carrier tracking loop. This loop attempts to drive phase errors with respect to the incoming RF signal to zero (i.e., It "locks on" to input signal). It is in the internal workings of the phase lock process where the Doppler and Carrier phase observables are generated.

There are multiple places in the signal processing where unwanted phase noise can creep into the observed signal phase. Some are noted in Fig. 10.8. By far and away, the largest of these disturbances to incoming phase is the quantization noise introduced in the 1-bit Analog to Digital Conversion. The sampling process in the Turbo Rogue corrupts the incoming signal phase with phase quantization noise, see Fig. 10.9. This noise is approximately uniformly distributed from -24.4 ns to $+24.4$ ns with a standard deviation of ~14 ns. The uniform distribution results from the non-commensurate sampling rate of 20.456 MHz with respect to incoming signal rates (in other words the sampling instant is moving in time as seen against the incoming signal). In addition, this phase quantization noise is wide band in its frequency content with an approximate bandwidth of $2 \times (20.456$ MHz$)$. For the reminder of this analysis, this will be assumed to be the dominant phase noise disturbance on the incoming signal and other sources of disturbance will be ignored. To be precise, we will assume that the sampling noise determines the receivers phase noise floor and hence its ultimate precision in time and distance.

Before proceeding with that analysis, it should be noted that the generation of the NCO-derived Carrier for down conversion to base-band will also have phase quantization noise (as noted above) with nearly the same uniform distribution as just described for the input sampling. This follows from that fact that even though the NCO has an *average* phase resolution of ~0.2 ps (for $L1$, a little less for $L2$), it will still have phase jumps of 48.8 ns when its internal accumulator overflows. It will be assumed that the mixing of these two quantization errors can be modeled as one source at the input to the carrier loop.

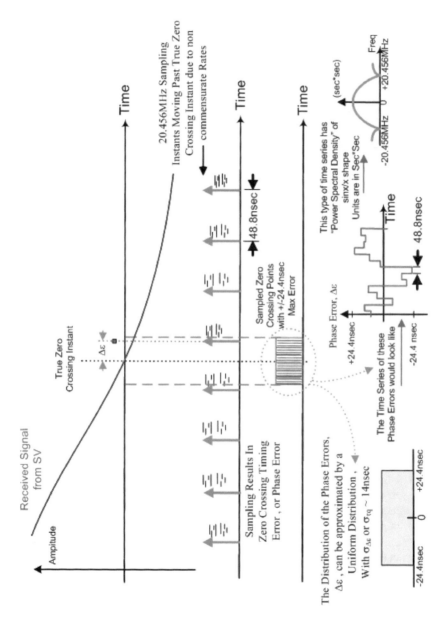

Fig. 10.9 The phase sampling process of an incoming sine-wave to model receiver

10.5.5   A Base-Band Model of the Carrier Tracking Loop with Phase Quantization Noise As the Input

Figure 10.10 shows a base-band model of the Turbo Rogue carrier-tracking loop. It is not really clear from the figure what part of this processing is done in hardware and what part is done in software. Up to the Integrate and Dump filter the signal processing is done in hardware. Also the NCO function is done in hardware. The phase difference is computed in SW by an approximation to the Arctan function. The inputs to the Arctan function are the accumulated I and Q data (HW). The accumulator clocks at 20.456 MHz and sums up the quantized input samples. Every 20 ms, the accumulator output is generated for use by carrier and code loops. At the 20 ms output time, the accumulator is cleared (or dumped) and the accumulation starts again. In this hybrid block, some functions are working at 20.456 MHz while others are working at 50 Hz. The actual Loop filter computations are done in software at a 50 Hz rate.

For this analysis of Carrier Phase measurement, we will assume the phase input to the receiver is zero or a constant. This leaves only the quantization noise as an input to the carrier-tracking loop, which produces Carrier Phase and Doppler estimates. By estimating the averaging effects of each stage of the signal processing, we can estimate the noise floor of the carrier phase measurements. To facilitate that analysis, a band-limited white noise approximation of the quantization phase noise is shown in Fig. 10.11.

In Fig. 10.11, the quantization phase noise is shown as a band-limited white noise power spectral density with amplitude 9.698×10^{-24} s $\times$ s. This approximation has the same mean square value as the uniform distribution discussed above (~14 ns). The Accumulate and Dump function has been approximated by a brick wall low pass filter with a bandwidth of 50 Hz. The same was done for the Loop bandwidth function. These approximations allow rapid estimation of the effects of the averaging of these two functions on the quantization noise. The low-pass filtering process reduces the noise power, which in turn reduces the mean square value of the noise. The final act of tracking the carrier performs an equivalent low-pass/average function on the injected quantization noise. In the end, we see that the quantization noise has been reduced to approximately a standard deviation of ~3 ps. This is the final estimate of the noise floor for the raw Carrier Phase and Doppler Observable under the assumption the quantization noise is the dominant internal phase disturbance. With this estimate, we can now say that our time precision is on the order of 3 ps for the raw Turbo Rogue Carrier Phase/Doppler observables.

Subsequent least square fitting of carrier phase observables was not considered here but that too is an averaging type function, and it would be expected to increase the time precision (or equivalently lower the noise floor of the carrier phase measurements).

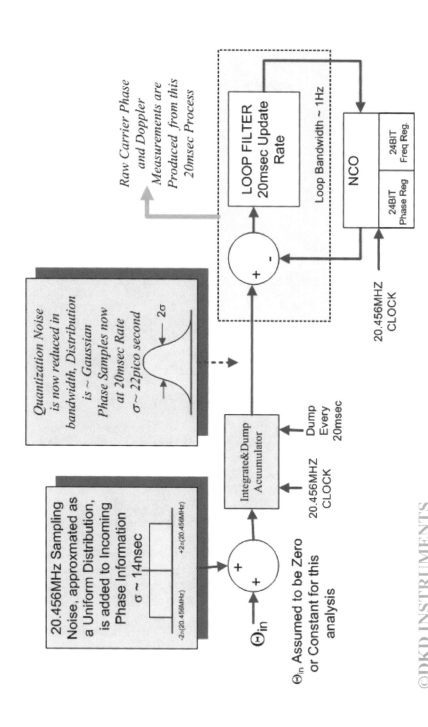

©DKD INSTRUMENTS

Fig. 10.10 Equivalent model/base-band carrier tracking loop with phase quantization error as input

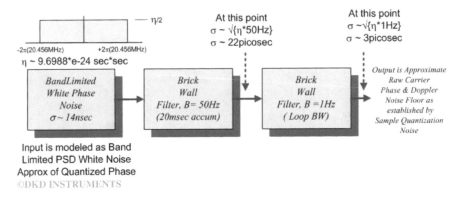

Input is modeled as Band
Limited PSD White Noise
Approx of Quantized Phase
©DKD INSTRUMENTS

Fig. 10.11 Equivalent white noise model input. Brick wall filter approximation

Table 10.2 Estimate of Incoming Timing Jitter using numerical SSB calculator, σ_{ti}, at the Input at Output of 20 Msec accumulator, Output of Carrier Tracking Loop L1 Carrier

C/No (dB)	BW (Hz)	σ_{ti} (pico-sec) (for L1 Freq)
40	10	6
50	10	1.8
60	10	0.6
40	50	18
50	50	6
60	50	1.6

10.5.6   *Estimates of Timing Jitter as Seen by Carrier Processing*

When a pure sine-wave plus noise is processed by the receiver, it will be observed to have phase jitter. The amount of jitter on the incoming signal is inversely proportional to the SNR of the received signal (thermal noise). Specifically, high SNR signals will have less jitter than low SNR signals. Predicting the exact level of this jitter is difficult. As an approximation, one can model the incoming process as a band-limited white noise signal plus the carrier signal. Such signals can be decomposed into quadrature components where the white noise power finds its way into the phase angle of the sine wave. Using this approximate model, a SSB phase noise to jitter calculator was used to estimate the timing jitter present on the incoming signal at the 20 ms Accumulator and the Carrier Tracking Loop, for typical GPS received carrier to noise ratios. The results of these calculations are shown in Table. 10.2 for BW's of 50 and 10 Hz.

Reference [1] details a formula that relates C/N$_0$ to RMS tracking error in degrees (Plot 10.2). If this tracking jitter is converted to equivalent timing jitter at the carrier frequency, the following formula results:

$$\sigma_{tL} = [1/2\pi f_c] \times \sqrt{\{[B_L/\alpha] \times [1 + 1/(2T\alpha)]\}} \qquad (10.23)$$

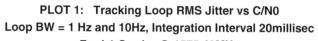

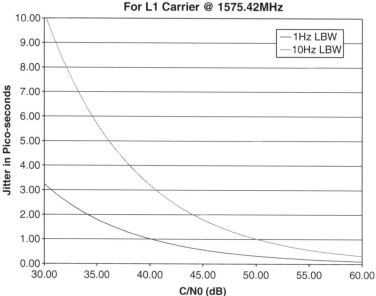

Plot 10.2   This plot shown below illustrates the estimated timing jitter for L1 against C/N_0 typical values with $B_L = 1$ Hz and 10 Hz, $T = 20$ ms, the internal integration gate period using (10.23) above

Where;

σ_{tL} is the timing jitter in seconds
f_c is the carrier frequency
B_L is the tracking Loop Bandwidth
α is the carrier to noise level, C/N_0
T is the pre-detection integration interval

10.5.7   *Summary Sampled Phase Analysis for Turbo Rogue Receiver*

The best time and distance precision of the Turbo Rogue receiver is contained in its Carrier Phase measurements. Estimates were presented to show the Turbo Rogue receiver has the ability to measure intervals of time to the precision of ~3 ps. It should be stressed that this time precision was estimated for raw 20 ms phase data with no curve fitting or post processing effects considered. In particular, the Turbo Rogue receiver time precision was estimated by considering the effects of phase quantization error, the modeled dominate internal phase disturbance on the carrier phase measurements with a final loop bandwidth of 1 Hz.

The incoming signal jitter was also considered and was found to have jitter in the ~0.2 to 3 ps level depending SNR for the 1 Hz carrier loop bandwidth, see Plot 10.1. Two methods were used to estimate this jitter, an SSB method (via numerical calculator) and (10.23) presented above. Equation (10.23) predicts lower jitter than the numerical result for a 10 Hz Loop BW.

It is expected that an aggregate of measurements of multiple SVs and post processing of raw 20 ms samples of carrier phase data will result in substantial increases in phase/time/range precision for the Turbo Rogue receiver. It seems reasonable from the analysis presented here to expect an average time precision from 20 ms Phase data of ~5 ps, or approximately 1.5 mm of range.

10.6   Summary

This chapter investigated a number of carrier phase topics: from the measurement of carrier phase and the extraction of Range information from these measurements to the limits on carrier phase measurements in general. In addition, information was presented that investigated the ratios and differences of integrated carrier phase. Much of the information used the Turbo Rogue receiver as the model but the material covered applies to many receivers with appropriate adjustments.

Reference

1. Braasch M, Van Dierendonck AJ (1999) GPS receiver architectures and measurements. In: Proceedings of the IEEE, p 59

Chapter 11
JPL Turbo Rogue Receivers

In this chapter, we examine the details of JPL designed Turbo Rogue Receiver. The Turbo Rogue receiver was funded by NASA and executed at JPL. The Turbo Rogue receiver is a dual frequency receiver capable of receiving both the L1 and the L2 GPS signals. It became one of the first commercially available receivers to use codeless techniques in the reception of the L2 signal. Initially produced as an eight channels of C/A, P1, P2 receiver, later versions contained L2 channels for C/A, P1 and P2 signals.

There are many public domain documents available that either directly or indirectly discuss the Turbo Rouge receiver. A good place to start is the US Patent office and the NASA Technical Reports web site (http://ntrs.nasa.gov).

11.1   Turbo Rogue L1 C/A Receiver

In this subsection, the Turbo Rogue receiver will be investigated. This receiver was designed by JPL under various US government contracts. There are a number of patents and other publications that are publicly available that detail the design and architecture of the Turbo Rogue receivers. Some the receivers introduced on the commercial market using Turbo Rogue (TR) methods were also able to provide codeless reception of the encrypted L2 signal as well the encrypted portion of the L1 signal. These receivers were some of the first commercial receivers (non military) that were able to use the L2 signal to measure the atmospheric delay by virtue of codeless operation.

The Turbo Rogue receivers achieved a level of precision that few receivers have been able to match. The exact reasons for this are complex but can be traced to extreme attention to detail in their design and the methods used to extract range delay estimates using code and carrier phase methods. Of course to achieve this performance, DSP methods were used for all base band processing in the Turbo Rogue receiver.

D. Doberstein, *Fundamentals of GPS Receivers: A Hardware Approach*,
DOI 10.1007/978-1-4614-0409-5_11, © Springer Science+Business Media, LLC 2012

In this subsection, we will cover some of the details of the Turbo Rogue L1 C/A processing. The processing of the L1 P signal and L2 P signal will not be discussed. Much of the P signal processing is similar to the C/A signal methods. It is hoped the writings here will help the reader to be able to get more from a reading of patents and other publications. In particular once the C/A processing is well understood it should help in understanding the P signal processing.

11.1.1   A Mechanical Clock Model of Turbo Rouge L1 C/A Channel Processing

Figure 11.1 shows a block diagram based on mechanical clock models of the Turbo Rogue L1 C/A receiver for a single channel. Typically there would be L2 C/A channels in a complete receiver. The TR receiver is quite different when compared against the Zarlink receiver discussed in Chap. 9. What is quite striking is how the TR measures Delay or Pseudo Range as compared to the Zarlink method.

Looking at Fig. 11.1, we can see that the replica clock dials above the 1 ms dial are not present. In the Zarlink, these dials exist as hardware, digital counters that are recorded at the TIC instant. In effect, these dials time scales have been taken over by software calculations in the TR receiver. The method for producing the Pseudo Range or Delay estimate in TR receiver is also a calculation done in software where in the Zarlink the pseudo range was captured from hardware registers at the TIC (or strobe) moment.

In fact the TR receiver produces its Pseudo Range estimates by a completely different method than many receivers that are based on recoding the phase state of the replica clock at a known time epoch. Instead the TR receiver makes only two actual phase measurements of its replica clock against the received clock and those are the Code phase error residual and the Carrier Phase error residual.

11.1.2   A Totally Coherent Design Using 20.456 MHz Master Oscillator

The TR receiver's precision reference clock is based on a master oscillator at 20.456 MHz. This clock signal is divided down to produce a 1 s timing mark that has ~48.8 ns phase steps. In terms of the IF input sampling, processing of these samples, phase residual measurements, time tags, etc, everything in the TR receiver happens on a 20.456 MHz clock tic. In short, the TR is a completely coherent and synchronous machine with the 20.456 MHz clock used as the *heart beat* of the receiver. Thus the very process of recording phase and rates of the replica clock at precisely known instants of reference clock time creates information on the SV to receiver delay.

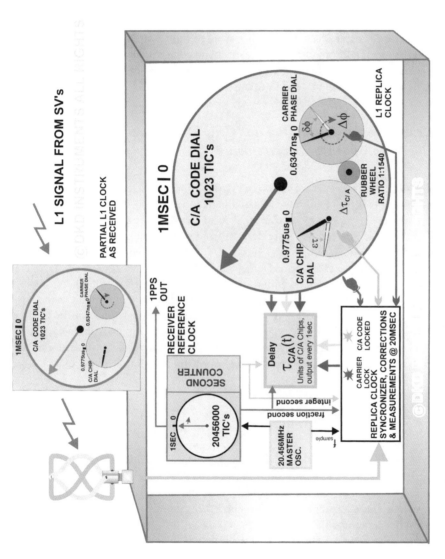

Fig. 11.1 Mechanical L1-C/A single channel clock model of turbo rogue receiver

11.1.3 Residual Carrier and Code Phase Measurements

There are two *fundamental* measurements the TR receiver makes, Carrier and Code phase residuals. The estimates of phase rate, delay, and replica clock control are directly or indirectly derived from these two measurements. The Carrier and Code phase residual measurements are shown in Fig. 11.1 as $\varepsilon\tau$ and $\delta\phi$, respectively. The $\varepsilon\tau$ measurement is the phase error of the replica C/A Chip Dial, against the received C/A Chip Phase Dial. The $\delta\phi$ measurement is the phase error of the replica clock Carrier Phase Dial with respect to the incoming Carrier phase information as received from the SV. The carrier and code residual phase measurements are performed every 20 ms. The carrier phase residual has units of cycles per 20 ms. The C/A Chip dial has units of C/A chips per 20 ms.

11.1.4 Carrier Phase Dial and Chip Dial, Phase Rate Measurements

These two measurements are shown in Fig. 11.1 as $\Delta\tau_{C/A}$ and $\Delta\phi$, respectively. The $\Delta\tau_{C/A}$ variable is the phase rate of the replica C/A Chip Dial. The $\Delta\phi$ variable is the phase rate of the replica clock Carrier Phase Dial. Both the Carrier phase rate and the Code Phase rate estimates are derived from measurements of the carrier phase residual, $\delta\phi$. By using the carrier phase rate to be rescaled as Code phase rate a much more accurate estimate of Code phase rate is achieved.

The two rates are specified with respect to the 20.456 MHz clock. Therefore any rate error on the 20.456 MHz clock will be present in these phase rate estimates. The two phase rate estimates are calculated every 20-ms. The units of the Carrier phase Dial rate are in cycles per 20 ms. The units of C/A Chip Dial phase rate are in C/A chips per 20-ms. The units are chosen to enable more direct updating of the digital oscillators used in the replica clock for Chip and Carrier phase dials.

11.1.5 Clock Synchronizer

The TR receiver uses only the code and carrier phase residual measurements as the basis for keeping the code and carrier loops locked. This task is accomplished in the Clock Synchronizer block of Fig. 11.1. The rate and phase of all three dials of the replica clock are under clock synchronizer control. In addition, the rate and phase information of those dials is passed on to the $\tau_{C/A}(t)$ block at known time instants (time tags) as established by the reference clock.

When the code and carrier dials of the replica clock are in close phase agreement with received code and carrier phase, the clock synchronizer has achieved code and carrier phase lock. The control loops that accomplish this task are updated every 20 ms.

11.1.6   Delay or Range Delay Estimate $\tau_{C/A}$

Using the two residual phase measurements, $\varepsilon\tau$ and $\delta\phi$, their precise time tags and the phase state of the replica clock, the TR receiver is able to determine an estimate of Delay $\tau_{C/A}(t)$. The units of measurement of $\tau_{C/A}(t)$ are in C/A chips. The delay term is an estimate of the total delay between the SV and the receiver. The delay estimate uses the receiver reference clock as a phase reference, which can contain a phase error with respect to GPS time. Therefore, if we rescale $\tau_{C/A}(t)$ by Code period (in seconds per Chip) and speed of light (Meters/Second), then we have an estimate of the Pseudo Range for the SV being tracked. The delay estimate is based on averages and curve fitting of past and present measurements with the final observable value reported every 1-s.

11.1.7   Code Rate Is Tied to Carrier Rate

The TR receiver uses a "carrier aided" C/A Code Tracking loop. This is indicated in model of Fig. 11.1 as the rubber wheel between the carrier dial and the C/A chip dial. In operation, the rubber wheel couples the carrier phase rate, $\Delta\phi$, to the C/A chip phase rate, $\Delta\tau_{C/A}$. The rubber wheel is scaled as a 1/1,540 reduction, the exact ratio between C/A chip dial and Carrier Phase Dial. As the dials are not gear driven, they can slip in phase but their rates remain nearly fixed at the correct ratio, i.e., Syntony. By using the carrier dial rate to drive the C/A chip dial rate, a smoothing action occurs on code phase measurements as the massive rate reduction is acting as a filter on any noise present on the carrier phase rate.

If the code loop were driven only by carrier phase rate, eventually the C/A code correlation would be lost. In the TR receiver, this is prevented by a highly averaged and slow feedback of the measured code phase residual, $\varepsilon\tau$, such that the C/A replica code remains aligned with the incoming code.

11.1.8   Why Is the Carrier Phase Rate, $\Delta\phi$, Subtracted from the Nominal IF Frequency to Obtain the Code Phase Rate $\Delta\tau_{C/A}$?

In Fig. 11.5, the details of the TR computations are detailed. Buried in those calculations is the equation that derives C/A code Rate from Carrier Phase Rate. As discussed above, the TR uses Carrier aided C/A code tracking. To arrive at the C/A code rate from the carrier rate the raw carrier rate needs to have several operations done on it;

- Sign Reversal of carrier Rate
- Add to 308 kHz, the Nominal IF
- Divide by the 1,540, the ratio Code rate to Carrier Rate

The sign reversal follows from the fact that for an approaching SV, the Doppler frequency is increasing, i.e., positive slope, while the Range delay is decreasing, negative slope. The 308 kHz must be subtracted as the TR carrier rate term has this bias rate inherent in it. The final divide by 1,540 takes into account the ratio between the two rates with all biases removed.

11.1.9   20 ms Update Rate

The 20 ms update rate used throughout the TR processing is chosen to coincide with the 50 Hz data rate present on the received C/A signal. When properly tracking, the TR receiver synchronizes the residual measurements, computations for code and carrier loops, and 20 ms type observables to the data bit transitions present on the SV signal being tracked. The time tags from the reference clock for each 20 ms interval are also ~ aligned with the data edges. The alignment is not perfect and can be off by a few microseconds typically. Periodic adjustment of the start time of the 20 ms interval is needed as it contains the scaled Doppler offset rate. In other words, the phase of received 20 ms data epochs are moving at the scaled Doppler rate against 20 ms epochs that could be generated from the reference clock.

By choosing to align its code and carrier loop processing with the 20 ms epoch of the received data, the TR avoids processing the incoming signal across a data bit boundary. This is advantageous as it avoids a sign change in the carrier control loop, which avoids using a Costas Loop type carrier tracker.

11.2   A Synthetic 20 ms Dial Mechanical Model and the Time Tag

The essence of the TR's use of a new time tag every 20 ms, that is coincident with the received 50 Hz data bit edges, is to create a *synthetic* 20 ms replica clock dial. We can form a mechanical model of the TR L1 C/A processing that includes the synthetic 20 ms and the hardware dials of C/A Code, C/A Chip, Carrier Phase and also the receiver reference clock. This model is shown in Fig. 11.2. In the model shown, we assume zero error in reference clock w.r.t GPS time, zero phase errors of replica clock dials including the synthetic 20 ms dial w.r.t received timing signals from the SV.

In our past use of this type of model (also see Zarlink), the phase state of the *replica clock* was typically captured by a strobe signal fired on the receiver reference clock's zero tic. The TR does it differently. Instead it mounts the strobe on the synthetic 20 ms replica dial and fires the strobe at zero tic, thus capturing the phase state of the *receiver's reference clock* at that instant. The reference clock instant recorded is the time tag for that strobe moment.

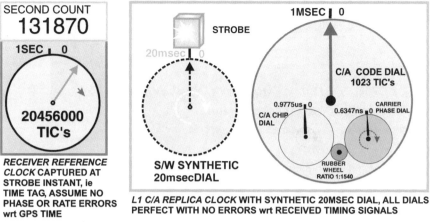

SECOND COUNT

131870

1SEC | 0

**20456000
TIC's**

*RECEIVER REFERENCE
CLOCK* CAPTURED AT
STROBE INSTANT, ie
TIME TAG, ASSUME NO
PHASE OR RATE ERRORS
wrt GPS TIME

time_tag =131870+ [2544720/2045600] sec

STROBE

20msec 0

**S/W SYNTHETIC
20msecDIAL**

1MSEC | 0

**C/A CODE DIAL
1023 TIC's**

0.9775us 0
C/A CHIP
DIAL

0.6347ns 0 CARRIER
PHASE DIAL

RUBBER
WHEEL
RATIO 1:1540

L1 C/A REPLICA CLOCK WITH SYNTHETIC 20MSEC DIAL, ALL DIALS
PERFECT WITH NO ERRORS wrt RECEIVED TIMING SIGNALS

Fig. 11.2 TR L1C/A mechanical model showing generation of time TAG from synthetic 20 ms
replica dial at strobe

The captured phase of the receivers' reference clock at the 20 ms strobe instant,
including the 1 s counter value, is exactly the time tag value used in the TR
calculations. The time tags will have a fractional component between zero and 1 s
with lLSB equal to 1/20.456 MHz and a whole second component.

If we assume the receiver's clock has zero error w.r.t GPS time, then the sub 1-s
component of the time tags has the *pure* range delay information in it, i.e., no receiver
clock bias would be present. This information is modulo 20 ms, i.e., we must subtract all
the whole 20 ms intervals from the sub second portion of time tag to obtain that portion
which directly pertains to range delay. This final fraction, $FRAC_{timetag}$ is ambiguous as
there are either 3×20 ms or 4×20 ms of time that must be added to it obtain the
complete range delay. This follows from the fact that the total delay to any given SV is
bounded by 60 ms at the minimum and less than 90 ms at the maximum.

Toward the end of the TR calculations (see below) for total Delay, a calculation
is performed to determine the start time of the C/A code generators, the variable C_k.
This calculation involves subtracting the current estimate of the Delay from the
current time tag plus an adjustment for C/A code rate. If we assume that the
placement of the time tag is ~ exact (i.e., exactly on *received* 20 ms time mark)
and the Delay has ~ zero error then this difference should produce a result that is
very near zero, *modulo 20 ms*. To be precise we would subtract all whole intervals
of 20 ms that fit in both the sub-second portion of the time tag and do the same for
the Delay *then* subtract. A ~ zero result just means we start the C/A code generator
at the beginning of its sequence with ~ zero phase offsets.

The actual TR calculations allow for the time tag to not be super precise. That is
it can be off a bit, typically less than a few microseconds. This small error is steered
toward zero by the once per second adjust of the time tag using the scaled difference

of two successive values of the computed delay $\tau_{c/a}$. The start time of the C/A code generation is of high precision as it uses the *difference* between delay, $\tau_{c/a}$, and time-tag, i.e., C_k.

11.2.1   Getting Delay Estimates Using the Time Tags from 20 ms Synthetic Dial with SV Clock as Reference Clock

In Fig. 11.3, the TR reference clock is shown as the 1 s dial with 20,456.00 tics and the integer seconds counter. We assumed the TR reference clock to have phase zero error with respect to GPS time. If we also assume the SV clock that we are tracking has zero phase errors w.r.t. to GPS time we can replace the TR reference clock with the SV Clock for recording time tag values. This is shown in Fig. 11.3. The second counter dial of the SV is omitted so the time tags taken will be sub second time tags. In addition the omission of the carrier phase dial of the SV clock means our time tags will be recorded as Code Phase type measurements.

As before in Fig. 11.3 all clock phases are captured at the strobe instant using light of infinite speed. The result is that the TR replica clock is showing *time sent* modulo 20 ms while the SV clock is showing *time received* with ambiguity of 20 ms. The absence of the 1 s dial in TR replica clock prohibits determining which phase of the reference clock 1 s dial (SV clock) is the actual receive time.

As shown in Fig. 11.3 six consecutive time tag phases are recorded by the receivers' synthetic 20 ms dial. As before, we will assume the 20 ms dial to be a perfect replica (zero phase error) of the received clock signals, as well as all the other dials in the replica clock. The resulting sub second time tags and derived delays are shown in Fig. 11.3. As can be seen from the values of the time tags and the computed delays, the SV is receding away from the receiver, i.e., the delay is growing.

The time tag derived delays show the result of stripping out all integer multiples of 20 ms, i.e., modulo 20 ms computation. The residual delay is now only known to this remainder phase residual plus either 60 or 80 ms. The ambiguity is due to the absence of a 1 s dial in the replica clock. To be precise, it is not known exactly which of the SV reference clock 1 s dial phases is the *correct* one. If the replica clock had a 1 s dial counting 20 ms epochs, we would know which occurred at the strobe instant. The phase difference between our missing 20 ms replica dial and the SV 20 ms dial would resolve this issue. This small ambiguity can be resolved in the navigation solution or perhaps as a new synthetic 1 s dial in the processing internal to the TR GP processor.

The model presented here has allowed us to examine how we can extract a modulo 20 ms estimate of range delay directly from the TR time tag. The accuracy of this method depends primarily on the accuracy of the replica dials being in phase with the incoming clock signals from the SV. The TR documentation states that the time tag placement or equivalently the synthetic 20 ms replica dial phase is not that accurate. In the patents and papers published regarding the TR receiver it is stated

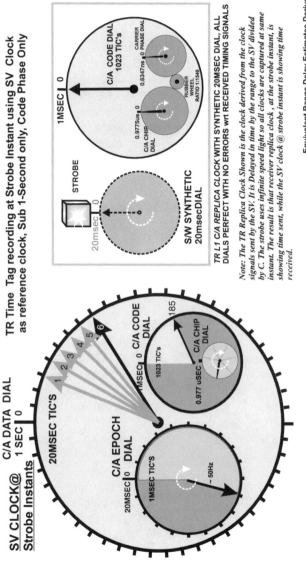

SV CLOCK @ C/A DATA DIAL **Strobe Instants** 1 SEC 0

TR Time Tag recording at Strobe Instant using SV Clock as reference clock, Sub 1-Second only, Code Phase Only

20MSEC TIC'S

C/A CODE DIAL
1023 TIC'S
1MSEC 0

C/A CHIP DIAL
0.977 uSEC

C/A EPOCH DIAL
20MSEC 0
1MSEC TIC'S
~ 50Hz
185

C/A CODE DIAL
1023 TIC's
0.9775us 0
C/A CHIP DIAL
1MSEC 0

CARRIER PHASE DIAL
0.6347ns 0

RUBBER WHEEL RATIO 1:1540

STROBE
20msec 0

S/W SYNTHETIC 20msecDIAL

TR L1 C/A REPLICA CLOCK WITH SYNTHETIC 20MSEC DIAL, ALL DIALS PERFECT WITH NO ERRORS wrt RECEIVED TIMING SIGNALS

Note: The TR Replica Clock Shown is the clock derived from the clock signals sent by the SV. It is Delayed in time by the range to the SV divided by C. The strobe uses infinite speed light so all clocks are captured at same instant. The result is that receiver replica clock , at the strobe instant, is showing time sent, while the SV clock @ strobe instant is showing time received.

Six Consecutive, Sub Second Time Tags recorded on SV clock @ strobe instant;

time_tag1 = 3*(20msec) + 11* (1msec) + 185* (0.977usec) + 0.05(0.977usec) =>   71.180745048msec
time_tag2 = 4*(20msec) + 11* (1msec) + 185* (0.977usec) + 0.183(0.977usec) => 91.180923791msec
time_tag3 = 5*(20msec) + 11* (1msec) + 185* (0.977usec) + 0.316(0.977usec) => 111.181053732msec
time_tag4 = 6*(20msec) + 11* (1msec) + 185* (0.977usec) + 0.459(0.977usec) => 131.181193443msec
time_tag5 = 7*(20msec) + 11* (1msec) + 185* (0.977usec) + 0.599(0.977usec) => 151.181330223msec
time_tag6 = 8*(20msec) + 11* (1msec) + 185* (0.977usec) + 0.739(0.977usec) => 171.181467003msec

Equivalent Range Delay Estimates Derived from time tag readings, Modulo 20msec, with 20msec ambiguity, N =3 or 4;

delay1 = 11.180745048msec + N*20msec
delay2 = 11.180923791msec + N*20msec
delay3 = 11.181053732msec + N*20msec
delay4 = 11.181193443msec + N*20msec
delay5 = 11.181330223msec + N*20msec
delay6 = 11.181467003msec + N*20msec

© DKD INSTRUMENTS

Fig. 11.3   TR time tag processing to obtain delay estimate

the time tag can be off by as much as a few microseconds. In such a case, the error must be known and accounted for or the resulting delay estimate derived from the time tag will also be in error by that amount. If this correction exists, it is not clear to the author where and how it is done.

11.3 Turbo Rogue Processor Block Diagram

The TR processing of the C/A signal is done using two fundamental blocks, which are shown in Fig. 11.4. The custom logic portion is for the high speed computations and operations associated with C/A code generation, Carrier Generation for conversion to baseband with I and Q paths, correlation of generated C/A code with received signal, accumulation of early, late, and prompt correlation sums and generation of a local receiver reference clock with 48.8 ns phase resolution. The final output of the hardware portion of the receiver is contained in the accumulators.

The General Purpose (GP) Processor is used to do the computations on the contents of the accumulators. Every 20 ms, the accumulators are read by the processor and the carrier phase and code phase residuals are computed. With the residuals and the time tag information, the processor can then solve for the next values for the control loops of code and carrier phase as well as the next delay.

11.3.1 Down Conversion and Sampling

The input signal is down-converted to an IF frequency of 308 kHz. The local oscillator(s) used for the down conversion are phase locked to the 20.456 MHz master oscillator. Any rate error on the 20.456 MHz will appear on the LO's locked to it. After conversion the input signal is split into I/Q components. The I/Q components of the input signal are then sampled by 20.456 MHz clock. The TR receiver can accommodate single bit (sign bit) or sign and magnitude signal samples.

11.3.2 The Reference Clock

The reference clock is composed of a 20.456 MHz oscillator along with a divider to produce the 1-s dial indicated in Fig. 11.4. The 1-s dial clock is used as a high-resolution timekeeper and also provides a 1PPS pulse. The second counting register also counts each 1 s pulse. Typically, the two clocks would be set to GPS time. The 1PPS dial being within 48.8 ns of GPS time while the second counter contains the seconds of the week information. The Reference Clock can be used by the GP processor to establish the precise instant the accumulators are started and stopped.

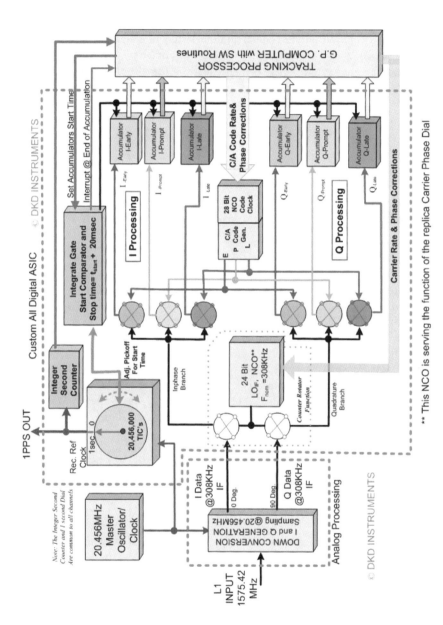

Fig. 11.4 TR block diagram, single L1 C/A channel

11.3.3 Accumulator Start/Stop Control, Processing

All the accumulators have a common start and stop time, which is determined by the GP processor as just noted. The value computed is loaded into a register, which acts as a pickoff point on the 1-s clock dial and a value for the integer seconds. When the time displayed on the 1-s dial and the 1-s counter equals that loaded by the GP processor a start pulse is sent to the accumulators. After 20 ms, the accumulators are stopped and their contents read by the GP processor, typically by an interrupt process.

The contents of the six accumulators now contain all the information needed to track the received SV signal and measure the delay to the SV being tracked. A 20-ms accumulation period will sum ~ 20 ms/48.8 ns or 409,120 input signal samples at 20.456 MHz sample rate.

11.3.4 Carrier Phase and Code and Chip Dials

The Carrier NCO, the C/A NCO, and the C/A code generator are the equivalent of the Carrier Phase Dial, C/A Chip Dial, and C/A code dial, respectively, of our mechanical clock model. The Carrier Phase and C/A Chip NCO's both are clocked at 20.456 MHz. The resolution in phase and rate of the Carrier NCO is 24 bits while the C/A chip NCO is 28 bits. The two NCOs can be controlled in Phase and rate by the GP processor.

The scaling of the rate phase information for the Carrier/Code NCO's as well as Code generator Phase is in cycles and C/A chips respectively. There is a final rescale of this information into counts just before it is loaded into the NCO or code generator register being commanded. For the carrier dial, the scale to counts is 2^{24} counts/cycle, for the C/A chip dial the scale is 2^{28} counts/Chip.

11.3.5 Absence of C/A Epoch Signal, 20 ms and 1-s Counters or Replica Dials

As can be seen in the block diagram, the C/A Epoch signal is not present nor is a 20 ms or 1-s Counter (Dial). As already noted above, the TR gets by without direct expression of these replica clock dials, which is quite unique. In some sense, the 20 ms dial is present as the Start/Stop time of the accumulators. But its jitter or timing uncertainty is greater than would be present on a 20-ms signal directly derived from the C/A epoch. In the computations for delay, $\tau_{C/A}(t)$, this time uncertainty or jitter is compensated for as the exact time of all events is recorded using the receiver reference clock.

11.3.6 C/A Code Generator

The details of the exact method of C/A code generation are not discernable from the public information available. But from what is said in the patents and reports covering the TR, it can be ascertained that a look-up table is used to initialize the code generator for a desired C/A code starting point. These g1/g2 values correspond to the start time contained in the computed C/A phase.

In addition only a small number of g1/g2 combinations are needed. This follows from the fact that the accumulators are started near the beginning of a *received* 50 Hz data bit. At the start of any 50 Hz data bit, the C/A code is starting from its beginning point as it cycles every 1 ms and 20 ms is an integer multiple of that period.

The look-up table must convert the whole, or integer portion, of the complete C/A phase command (in units of C/A chips) into g1/g2 values. The fractional part of this phase command is sent to C/A Chip NCO phase input.

11.3.7 Carrier Phase NCO Command Is Extracted from Total Integrated Carrier Phase

In the TR receiver, the calculated command for the Carrier Phase NCO is phase contained in a variable that has the *total* number of carrier cycles at nominal IF rate of 308 kHz that have transpired since the start of the SV track, i.e., *Total Integrated Carrier Phase*. Therefore it has a whole cycle count plus a fraction cycle count. The fraction of a cycle is what the Carrier NCO is loaded with at the start of any accumulation interval.

If we assume zero rate error on the 20.456 MHz clock (i.e., Zero Doppler input is at 308 kHz exactly) and the initial total phase $\Phi(t_0)$ is set to zero at the start of a track, time t_0, then the total Phase, in cycles at any time t, after start time t_0 is;

$$\Phi(t) = [t - t_0][308 \text{ kHz}] + \int_{t_0}^{t} f_d(t) \, dtZ$$

where $f_d(t)$ is the SV Doppler Rate versus time

A discrete version of the above calculation is at work in the TR receiver and its units are in cycles with whole cycles and fractional cycles accounted for. Only the fractional part of the total integrated phase is sent to the Carrier NCO phase input point. Note the presence of the 308 kHz term in the above equation, this is the nominal LO_{if} (308 kHz) rate as used by the TR to down-convert the sampled input signal to base-band (DC).

11.3.8 Phase and Rate Steered Carrier Phase Loop

The TR receiver commands both rate and phase of the Carrier Phase NCO (Dial) to maintain phase lock on the incoming signal. The loading of the Carrier Phase NCO phase at the start of any 20 ms interval is a departure from what many other receivers do, that is they are rate only steered carrier loops. The method used follows the JB Thomas paper on Digital Phase Locked Loops. The result is carrier NCO phase that is not continuous; it can have jumps of Phase. This contrasts sharply with the rate-only methods used by many other receivers.

11.3.9 Time Tag Information

The GP processor can get time from the Receivers Reference clock as a fraction of a second and as an integer second. The integer second counter would typically be set to the GPS seconds of the week count. The adjustable tap point on the 1-s dial sets the fractional 1-s of the start time for the accumulation. A comparator on the second counter does the same operation. In this fashion, the time tag for the current accumulation is established.

11.4 Turbo Rogue Receiver DSP Calculations, Performance Overview

The TR receiver DSP architecture is an exceptionally well-executed design that produces very accurate range estimates in both C/A and P codes. This accuracy can be traced to methods used in control, measurement, and computations integral to combined hardware DSP and software DSP that is executed in the GP processor. In summary, here are some the reasons the high performance is achieved;

- The goal of the TR design is to produce the most accurate estimates of Delay, Carrier Phase, Carrier Phase Rate, and Carrier Phase Rate-Rate observables possible at 1 s reporting intervals using only the *measurements* of Carrier Phase Residual and Code Phase residual for any given channel. The methods used to extract these two fundamental measurements ensure great fidelity. Since all estimates of observables follow from these two measurements, high fidelity is achieved for the derived rates and phases.
- The digital carrier tracking loop uses a mid-point method of phase estimation (model phase method) and commands phase as well as rate estimates. It is reported that this method produces a lower mean square phase estimate compared to traditional rate only, non mid point, digital phase lock loop methods.

- The addition of phase error commands to the carrier loop allows for reduced phase tracking errors. In particular, the classic second order, rate only, loop suffers from non-zero steady state phase error when the input *rate* changes with time, i.e., ~steady , non-zero acceleration (the normal case for GPS SV tracking). With phase feedback, the bandwidth of the carrier loop can be made narrower and not incur the increase in steady state phase error as would be the case in a second order rate only carrier loop. Note: Nearly all linear models of PLL first, second, and third order loops under dynamic stress (as seen in many texts) are for *rate only* steered loops.
- The residual carrier estimate is used to derive the carrier phase rate and the carrier phase every 20 ms.
- The Carrier Phase rate is used to drive the C/A code rate, i.e., Carrier Aided C/A code tracking.
- The 20 ms carrier phase residual estimates are "averaged" over a 1-s interval by a curve fitting via least squares. The result is a set of observable's Carrier Phase, Carrier Phase Rate, and Carrier Phase Rate–Rate, centered on the GPS 1 s timing mark that is the best fit of measured residuals and their respective time tags.
- Integrated Carrier Phase is now a highly averaged measurement. Compare against the Zarlink method, which takes only one measurement and TIC instant (0.1–1 s interval)
- The code residual is averaged over 1 s and introduced as a low rate phase feedback to keep the code-tracking loop from drifting away in phase.
- The averaging on the code residual reduces the noise for the Code Phase measurements.
- The integration of the precise time tags and the accumulator start–stop times allows the TR to estimate the Delay without using an explicit 20 ms and 1 s dials.
- The weight of the above methods is to reduce noise in the observables of Delay, Carrier Phase, Carrier Phase Rate, and Carrier Phase Rate-Rate. In addition, the final observables are projected to the receiver reference clock 1 s epoch. As the navigation solution steadies out, the observables will be now be within a known offset to GPS 1 s epoch/time. That offset is known within the accuracy limits of the receiver's solution for receiver clock bias, typically a few 10's of ns.

11.5 Details of DSP Computations Performed in the GP Processor

The DSP computations done in the GP processor are responsible for delivering the observables of Delay, Carrier Phase, Carrier Phase Rate (Doppler), and Carrier Phase Rate–Rate. As noted above these 1-s observables come from the two measurements of Code and Carrier residuals. This remarkable fact comes with a price: Computational complexity.

In Fig. 11.5, the computations are traced from their beginnings as outputs from the six accumulators through to the final outputs of the 1-s observables just noted. The calculations shown are a *simplified* version of what is presented in the patents and papers relating to the TR receiver. Some steps were simplified, some presented as a block, and some omitted. But most of the *fabric and direction of calculation flow* of just how it is done is in retained Fig. 11.5. Here the emphasis is on flow of processing, scaling, and physical meaning. For the calculations of delay, the calculations presented follow those presented in JPL publication 88-15. The carrier loop calculations follow more closely those used in the JPL publication 89-2, Digital Phase Lock Loops, J.B. Thomas. Specifically the use of the second order loop filter and its gain constants. In summary, here are the differences and notes for the calculations presented here;

- Two fundamental calculation periods are present, 20 ms and 1 s periods
- The subscript index used in publication 88–15 for the 20 ms interval is n. The subscript k is used here. The subscript k is meant to be ~ centered on the 1 s epoch from the receivers' reference clock. On average, there should be 25 values of k on each side the 1-s epoch, or 50 total. This rule applies for the sums that add up 20 ms data to create a 1 s data point. In practice the 50 total can be plus or minus 1 depending data bit re-sync issues of the time tag.
- Here the subscript n is used to denote 1-s interval computations.
- The complication of a dead period is omitted. Specifically the accumulation interval, Δt, is assumed to be exactly 20 ms throughout the calculations with no time allowed for hardware update.
- The final steps needed to prepare the phase and rate feedback information to load into the Carrier NCO were omitted.
- A 308 kHz IF frequency was used, as was a 20.456 MHz sample clock rate. This is in step with later versions of the TR receiver, which also used full quadrature for P signal processing.
- The quadratic curve fit of the carrier phase data to get final observables of Carrier, carrier Rate, and Rate–Rate were omitted.
- The post 1-s curve fitting, i.e., the N-second methods were omitted.
- The indexing was simplified for readability. The exact interval that variables were tied to at certain points in the calculations was allowed to stray from strict to make it more readable.
- The portions that pertained just to P code were omitted.
- The mid interval time tag is not shown and the sums of the mid interval time tags were omitted.
- A different discriminator was used to calculate the code phase residual. Also a new variable was used, $\varepsilon\tau$, to represent the new non-coherent dot product discriminator output. In addition the integration of the 20 ms $\varepsilon\tau$ data to arrive at 1-s $\delta\tau$ data was performed in a slightly different fashion. The reason for these changes was to introduce a 20 ms code phase residual at roughly the same place in the calculation chain as the 20 ms $\delta\phi$ carrier phase residual. Lastly, a different name was used for the integrated residual gain, $G_{c/a}$ instead of Kc.

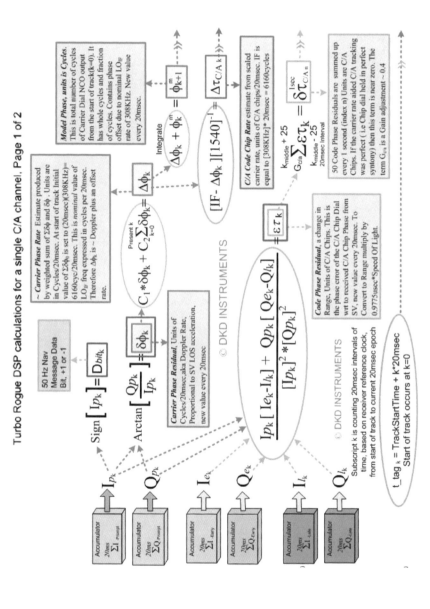

Fig. 11.5 Turbo rogue DSP calculations

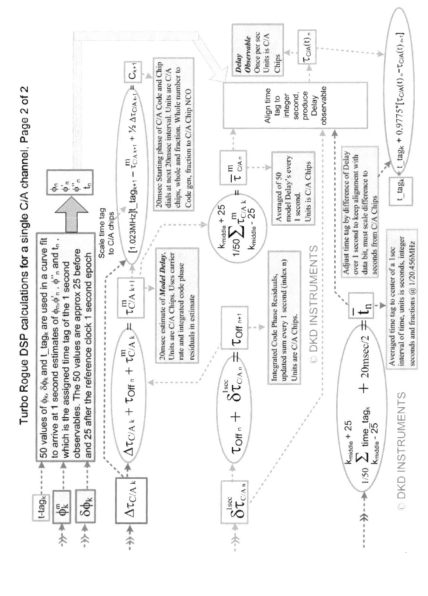

Fig. 11.5 (continued)

- In the step to calculate C_k, the starting phase of the C/A code generator in C/A chips, the time-tag term is introduced. The estimate of delay is subtracted from the time-tag and ½ of the C/A code rate per 20 ms added (all units of C/A chips). The C_k calculation is feeding back the *difference* (or error) between the time tag and Delay such that the *next* C/A starting point will reflect some of that error. The time-tag *error in time placement* corrections are done once per second by applying the time difference between two successive estimates of the total delay $\tau_{c/a}$, rescaled to units of seconds, see Fig. 11.5. In summary, the time tag is not static, but must be adjusted, as referenced to the receivers reference clock, such that the start time of the accumulators is approximately at the received instant of the 50 Hz data bit edge. This follows directly from the idea of the time tag following the synthetic 20 ms dial phase. That phase moves against GPS time as seen at the receiver reference plane due to the Doppler present on the receiver SV signal.

11.6   The Counter Rotator Term

In the papers and patents written about the TR receiver is a term that is called *Counter Rotator* or CR for short. This term was new to the author and I will set here what I believe are the reasons behind this terminology.

The term CR refers to a subsystem of the TR receiver, see Fig. 11.4. The subsystem it references is the NCO *and* the quadrature multiplication of the NCO output with the received signal from the SV. If the incoming signal is modeled as analytic signal at rate ω and the NCO is *locked* to it at rate $-\omega$ then the resulting quadrature multiplication is;

$$e^{jwt} \times e^{-jwt} \geq e^{0} \geq 1$$

There is a mechanical clock interpretation of the above equation. The equation could represent a received clock and a replica clock both rotating at rate ω, but the replica clock is rotating in the opposite direction, or *counter rotation*, of the received clock. If we use these two clocks as inputs to a differential gear mechanism, a mechanical quadrature multiplier, the output shaft would be stationary. The stationary output is the direct reflection of the result of the above equation, i.e., no rotating terms are left, just the unity constant, or DC term.

11.7 A Summary of TR Time Tag and the Delay ($\tau_{c/a}$) Observables

The TR generates the observables of time tag and $\tau_{c/a}$ in a complex fashion when compared, for example, to the Zarlink. Some of the details of the exact nature of these two observables are not documented. What is expressed here is the author's best understanding of these two terms:

- The time tag and the delay, $\tau_{c/a}$, both contain the range delay information
- The time tag estimate of Range delay is a low precision estimate.
- The Delay estimate of Range delay is a high precision estimate.
- If there is a phase error on the receiver clock, T_{bias}, the time tag and delay terms will contain this same phase error.
- The delay term, $\tau_{c/a}$, is units of C/A chips and is a direct measure of the complete Range delay to the SV being tracked. In other words, it is in the range of 60–80 ms when scaled in units of seconds. If the receiver clock bias is zero, then it is an unbiased estimate of the delay to the SV being tracked. If we multiply by seconds per C/A chip and then by speed of light we get the range to SV. If the receiver clock bias is non-zero, then it is a pseudo range to the SV.
- If the receiver clock bias is zero then the time tag is an estimate of the *time received* modulo 20 ms as referenced to GPS time. The time sent is the corresponding 20 ms epoch at the SV.
- Therefore the time tag contains range information to the SV that is modulo 20 ms. In other words, one must subtract from the time tag all the integer multiples of 20 ms that fit into it. The remaining residual is a modulo 20 ms range delay estimate. To get the full range, you must add either 3×20 ms or 4×20 ms to this residual to obtain the final range delay between 60 and 80 ms.
- The time tag is adjusted, once per second, with information derived from the delay, $\tau_{c/a}$, so as to move it in relation to receiver reference clock time. This ensures the time tag is near the SV 20 ms data bit edge.
- In contrast, the delay, $\tau_{c/a}$, is adjusted every 20 ms with a new value as derived from carrier phase residual, i.e., carrier aided tracking.
- During the time between time tag adjustments, the data bit edge will move. Therefore during any 1 s interval, the error of the time tag is typically growing with respect to its *true* position and w.r.t to $\tau_{c/a}$. The error growth rate is proportional to the Doppler during that particular 1-s interval.
- What this means is that at the start of any 1-s interval, the time tag and delay, $\tau_{c/a}$, could have zero difference between them (modulo 20 ms). But as that 1-s interval draws to close, with 50 adjustments of $\tau_{c/a}$ been done, the error between the two will have grown by an amount proportional to the Doppler rate over 1 s of time.
- If the receiver is in stable tracking, has a near-perfect reference clock (an atomic one) set to GPS time (i.e., $T_{\text{bias}} = 0$) and the SV being tracked has a nearly zero Doppler value, the author would expect nearly identical estimates of range delay from both $\tau_{c/a}$ and the time tag (when expressed modulo 20 ms) both over the sub second interval and intervals above the 1 s interval.

11.8   Summary

This chapter was devoted to presenting some of the details of the Turbo Rogue receiver, specifically L1 C/A channel processing. As stated at the beginning of this chapter, few receivers have matched the TR in precision. The information presented here was incomplete in that it did not touch on the P code processing, the codeless processing, or the L2 channel that the TR receiver is capable of receiving. Rather the focus was on L1 C/A channel processing, which is inline with the goal of this text. Still significant material was presented and it is hoped for those studying this receiver that it helps them understand the portions not discussed.

Chapter 12
The L2C Signal

Contributed by Danilo Llanes, GNSS ITT

12.1 Introduction

This chapter introduces the L2C GPS signal. The first section of this chapter provides a brief history of the L2C signal, a list of the current satellites with L2C capability, a description of the enhancements available with L2C, and a discussion of the issues that drove the L2C design. It is followed by a section that provides a summary of the signals transmitted on L2 and a detailed description of the L2C signal waveform. The chapter continues with a section that describes how to demodulate and extract navigation data. Finally, it ends with a section that describes how to calculate the pseudorange, phase, and doppler observables.

12.2 L2C Signal History

In 1998, a White House press release announced the addition of a civil signal in the L2 (Link 2) frequency (1,227.6 MHz). It is now referred to as the L2 civil (L2C) GPS signal. The L2C signal is the first of the modernized civil GPS signals to be deployed in the GPS constellation. It was designed as part of the effort to meet the requirements and demands for GPS in the twenty-first century and is not transmitted by the legacy (or original) GPS satellite constellation which are comprised of Block II/IIA/IIR satellites currently orbiting the Earth. L2C provides greater position accuracy and faster signal acquisition than the current L1 C/A signal available to civilian users.

The first GPS satellite launched with L2C capability, Satellite Vehicle (SV) 53 transmitting Pseudo Random Noise (PRN) 17, occurred in September 2005. In December 2005, SV 53/PRN 17 was set to an operational (healthy) state. Additional L2C capable satellites have been and will continue to be incrementally phased into the GPS constellation as older satellites need to be replaced. Therefore, L2C is of limited use until the signal is broadcast from 24 satellites. It is expected that L2C

D. Doberstein, *Fundamentals of GPS Receivers: A Hardware Approach*,
DOI 10.1007/978-1-4614-0409-5_12, © Springer Science+Business Media, LLC 2012

Table 12.1 L2C satellites in
GPS constellation

Date set to an operational state	SV	PRN
December 2005	53	17
October 2006	52	31
December 2006	58	12
October 2007	55	15
January 2008	57	29
March 2008	48	7
March 2009 (launched)	49	1
August 2009	50	5
August 2010	62	25

will be available from 24 satellites by approximately 2016. As of May 2011, nine GPS satellites have become operational that transmit the L2C signal as indicated in Table 12.1.

When first launched, L2C satellites did not transmit navigation data messages. Instead, the satellites only transmitted the L2C PRN codes without any navigation data modulation. L2C navigation data was switched off until 2010 when it was enabled in order to support civilian user equipment testing. The current L2C navigation data message being broadcast is the default navigation message, referred to as Message Type 0. The default navigation message contains a properly structured message with an alternating pattern of logic ones and zeros in place of the navigation information that is contained in a normal message. It does not contain the full navigation message. L2C broadcasts will continue to transmit the default navigation message until the GPS Operational Control Segment (OCX) becomes fully operational. As part of the GPS modernization program, OCX will provide new command and control support capabilities for current Block II and future Block III GPS satellites including improved security, accuracy, reliability, and anti-jamming capabilities.

12.2.1   Applications for L2C

L2C was designed to specifically meet the commercial needs of civilian users of GPS and is the second GPS signal that is available to civilian users. The first civilian GPS signal was C/A (Coarse Acquisition) transmitted in the L1 (Link 1) frequency (1,575.42 MHz). Although L2C signals were intended to be available for all civilian GPS applications, the design was driven by two primary considerations:

1. Provide enhanced single frequency GPS applications such as

   a. Positioning inside of buildings
   b. Navigation in wooded areas
   c. Vehicle navigation in tree lined roads
   d. GPS in wireless phones

2. Support dual frequency civilian user applications such as

   a. Scientific applications to monitor geographic activities (earthquakes, volcanoes, and continental drift)
   b. Cadastral and construction land surveys
   c. Guidance and control of mining, construction, and agricultural machinery

12.2.1.1 Single Frequency Applications

An improved capability for a system is generally necessary to address a shortcoming in the design of the previous generation of that system. That is the case for L2C (an improvement) when compared against L1 C/A (the first generation civilian signal). In order to enhance single frequency applications for civilian users employing L1 C/A, L2C had to overcome the cross-correlation interference inherent with the L1 C/A signal and had to lower the data modulation detection threshold which allows decoding navigation messages when the receiver is barely tracking.

Cross-Correlation Interference Improvement

To explain cross-correlation interference and the reason that it became an issue for single frequency users, it is necessary to discuss GPS signal correlation. Correlation is a measure of the similarity between two GPS satellite signals. For GPS, all of the satellite signals are modulated on the same carrier frequency using a communications technique called Code Division Multiple Access (CDMA). For this discussion, it is the L1 frequency. In order to separate or distinguish each individual satellite's C/A signal, each satellite modulates a unique Pseudo Random Noise (PRN) sequence known as a Gold code onto the carrier frequency. There are 32 unique satellite PRN sequences which are designated as PRN 1–32. Gold codes appear to be random binary sequences but are actually deterministic sequences (hence the term pseudo random). After demodulating the carrier frequency, the signal is decoded by performing an auto-correlation (correlation of a signal with itself) of the received PRN sequence with a replica of the desired PRN sequence. At this point in the signal processing is where the interference enters the system.

The Gold codes are designed such that one satellite signal may be easily distinguished from another satellite (or are referred to as being highly mutually orthogonal) and are designed to have good auto-correlation properties. Therefore, the Gold codes are not supposed to interfere with one another, and a receiver should not be mistaking one satellite with another. However, cross-correlation interference occurs when an undesired satellite's signal is mistaken for the desired satellite's signal. Cross-correlation interference is defined as a sufficient amount of noise created in a receiver channel by an undesired satellite's signal which triggers detections, causes the wrong PRN to be temporarily acquired, and/or delays the acquisition of the desired PRN code. This type of interference is typically caused when strong C/A satellite signals interfere with weak C/A signals. The strong signals are typically due to the satellite being closer in range to the receiver or the receiver having an unobstructed view of the satellite. Weak signals are typically caused when the satellite is furthest from the receiver (at low elevation angles) or when obstructions, such as trees, interfere with reception. This cross-correlation interference effect is an inherent aspect of L1 C/A signals, and it is caused by the properties of the Gold codes.

The objective of the L2C design was to eliminate or greatly reduce the L1 C/A cross-correlation interference issue. The PRN codes selected for the L2C signal's worst case cross-correlation performance are measured to be 45 dB while the equivalent measurement for the L1 C/A signal is 21 dB. Therefore, the L2C PRN codes were designed to substantially reduce cross-correlation interference (by about 250 times).

Lower Data Modulation Detection Threshold

L2C was designed to provide another enhancement to single frequency civilian users of L1 C/A. The data modulation detection threshold is lower, yielding greater data recovery, and slightly better carrier tracking even though its transmitted power is lower than L1 C/A. Two specific enhancements to L2C provide this improvement over L1 C/A:

1. A pilot carrier signal
2. Forward Error Correction (FEC) encoding of navigation data

A pilot carrier signal is a waveform that has a PRN code modulated on the carrier without any navigation data modulation (dataless). In L2C, it is transmitted along with another waveform that contains navigation data modulation. The dataless signal is easier to acquire than the signal encoded with navigation data. After the dataless signal is acquired, then the signal with navigation data is acquired. This method improves signal acquisition and increases power levels at the correlators.

In L2C, the signal that contains navigation data modulation is FEC encoded (convolutional coding). Legacy L1 C/A navigation data was not encoded. Any interruptions due to interference or weak signals would result in losing navigation data, especially because the navigation data transmission rate is slow (50 bits/s). FEC encoding allows the receiver to detect and correct navigation data message errors. This enhancement provides a considerable improvement to navigation data recovery in conditions where the receiver is barely tracking.

12.2.1.2 Dual Frequency Applications

Dual frequency applications refer to receivers that use the L1 and L2 frequencies to perform GPS measurements. Using both frequencies allows the receiver to calculate ionospheric delay errors and correct the observables that are measured.

This technique permits greater position accuracy to be calculated. For receivers that do not use dual frequencies, a model must be used to calculate ionospheric errors or corrections from an external source must be obtained. As a consequence, dual frequency civilian users are generally professional or scientific users which require higher performance from their GPS receivers.

Prior to the availability of L2C, L1 C/A could be used by civilian users, but a civil signal on L2 did not exist. Dual frequency civilian users had to perform semi-codeless tracking techniques on L2 using the Precision PRN code, P(Y) code, reserved for military applications. L2C eliminates the need to perform semi-codeless tracking techniques with L2 P(Y) code. Therefore, civilians with dual frequency GPS receivers may now expect to use L1 C/A with L2C and obtain better accuracy.

12.3   L2 Signal Requirements

Prior to describing the details of L2C GPS signals, the signals that are required to be transmitted on L2 must be defined. GPS satellite signals that are transmitted on the L2 frequency are not the same for all satellites in the constellation. The signals that are transmitted depend on the type of GPS satellite. The term "Block" is used to designate the type of GPS satellite. The original set of GPS SVs was designated as Block I. In 1995, all of the Block I satellites were decommissioned. The designation for the current GPS operational satellites are shown in Table 12.2 as well as whether it is capable of transmitting L2C.

As Table 12.2 indicates, Block IIR-M, IIF, and III are the only operational satellites that are capable of transmitting the L2C signal. However, the requirements for the L2C signal for Block IIR-M satellites depend on whether L2C Initial Operational Capability (IOC) has been achieved. The L2C signal transmitted by Block IIF and III satellites does not depend on IOC. Table 12.3 shows the signals transmitted on the L2 carrier as a function of satellite block.

As the footnote in the Block IIR-M row of Table 12.3 implies, the L2C signal that may be possibly transmitted by a satellite is subject to change prior to IOC in approximately 2016. For example, PRN 17 initially transmitted the CM/CL PRN code without navigation data modulation from the time it was set operational in 2005 until 2010 where it started transmitting the default navigation message.

Table 12.2 Operational SV blocks

SV block	SV number	L2C capable
II	13–21	No
IIA	22–40	
IIR	41–61	
IIR-M		Yes
IIF	62–73	
III	74–81	

Table 12.3 L2C Transmissions versus SV blocks

SV block		L2 (1,227.6 MHz) In phase			Quadrature phase		
II	PRN code	P(Y)	P(Y)	C/A	Not available (N/A)		
IIA	Data type	NAV	None	NAV			
IIR	Data rate	50 bps		50 bps			
IIR-M[a]	PRN code	P(Y)	P(Y)	CM/CL	CM/CL	C/A	C/A
	Data type	NAV	None	NAV/None	NAV/None	NAV	None
	Data rate	50 bps		50 bps/None	25 bps/None	50 bps	
	FEC	N/A	N/A	No/No	Yes/No	No	No
IIR-M	PRN code	P(Y)	P(Y)	CM/CL	C/A	C/A	
IIF	Data type	NAV	None	CNAV/None	NAV	None	
III	Data rate	50 bps		25 bps/None	50 bps		
	FEC	N/A	N/A	Yes/No	No	No	

[a] This possible signal configuration only applies prior to the Initial Operational Capability (IOC) of L2C

Table 12.4 GPS carrier frequencies

Link	Frequency (MHz)	Multiplier	Application
L1	1,575.42	154	Civilian/Military
L2	1,227.6	120	Civilian/Military
L3	1,381.05	135	Nuclear detonation detection
L4	1,841.4	180	Study ionospheric correction
L5	1,176.45	115	Civilian safety of life
L6	1,278.75	125	Distress alerting satellite system

12.3.1 L2 Signal Structure

GPS satellite carrier frequencies reside in the L-band range of frequencies, from 1 to 2 GHz. To generate the various carrier frequencies used by GPS, the atomic clocks onboard the satellites (either a Cesium or Rubidium atomic clock) generate the fundamental GPS frequency 10.23 MHz. Each GPS satellite coherently derives all transmitted signal components, carriers, PRN codes, and navigation data, from the same onboard atomic clock fundamental frequency reference. All GPS carrier frequencies are generated by multiplying up the fundamental frequency by a multiplier value. For L2, the multiplier value is 120. Therefore, the L2 frequency is centered at 1,227.6 MHz. A list of all the GPS carrier frequencies is provided in Table 12.4.

The L2 carrier frequency is composed of two signal components that are in phase quadrature with each other or 90° out of phase. One signal component, P(Y), is defined as being transmitted in phase while the other signal component, L2C, is defined as being transmitted in quadrature with P(Y).

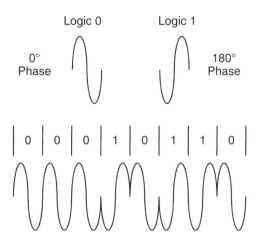

Fig. 12.1 Bi-phase shift keying (BPSK)

Table 12.5 L2 Range code combinations

Option	In phase	Quadrature phase (L2C)
1	P(Y)	CM/CL
2	P(Y)	C/A

Although up to four different ranging codes may be modulated on the L2 carrier, P(Y), CM, CL, and C/A, the options that are defined in Table 12.3 limit the L2C range codes and have been simplified to the following combinations:

The L2 carrier component that is defined to be in phase is the P(Y) ranging code used by the military. The P(Y) ranging code is modulated onto the L2 carrier and is a 10.23 MHz signal that employs bi-phase shift keying (BPSK). BPSK is a digital modulation technique that represents data, such as a binary data with two logic states, by changing the phase of the carrier signal between two phases that are separated by 180°. Figure 12.1 provides an example of a byte of binary data being converted to BPSK modulation.

The other L2 carrier component that is in quadrature with P(Y) is the L2C ranging code, either one of the two possible combinations of ranging codes defined in Table 12.5.

12.3.2   L2C Signal Description

The L2C ranging code, CM/CL, or C/A is modulated on the L2 frequency in quadrature with the P(Y) ranging code. Note that CM and CL are separate ranging codes that are always transmitted together, hence the term CM/CL. The CM/CL and C/A ranging codes are never transmitted simultaneously. Therefore, either CM/CL or C/A is transmitted. Figure 12.2 depicts the possible combinations of L2C signals that may be transmitted by a GPS satellite.

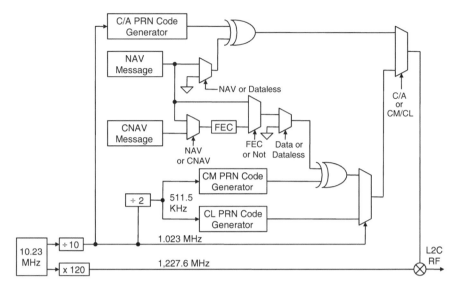

Fig. 12.2 Combinations of L2C Signals

The choice of whether to modulate CM/CL or C/A on L2 is selectable via commands from the GPS control segment. The CM/CL and C/A ranging codes are both 1.023 MHz signals that use BPSK.

As noted earlier, the CM and CL ranging codes are transmitted together. However, the sequences are not transmitted simultaneously. As depicted in Fig. 12.2, the individual chips of the CM and CL sequences are time multiplexed at a 1.023 MHz rate. Figure 12.3 provides an example of the interleaving of CM and CL chips into a composite 1.023 MHz signal.

12.3.2.1   Civil Moderate Ranging Code

The Civil Moderate (CM) ranging code is one of the PRN codes available in L2C. The CM PRN sequence is generated as a linear binary pattern that repeats every 20 ms. The code chip rate is 511,500 Hz. A chip is a single cycle of the PRN code sequence. Therefore, the code length is 10,230 chips (511,500 chips/s × 0.020 s).

The CM PRN sequence is generated in hardware with a shift register and exclusive-OR gates. A shift register is a chain of flip-flops that transfer the input binary data stream through the encoder. The exclusive-OR gates (depicted as a circle with a plus symbol) perform modulo-two addition of the inputs and are positioned at locations defined by a generator polynomial. The CM PRN code generator uses a 27° generator polynomial (hence a 27 bit shift register) defined in octal notation as 1112225171_8. The shift register is clocked at a rate of 511,500 chips/s. It is reset every 10,230 chips to an initial value stored in the CM PRN code initial state register. Each satellite has a unique CM PRN code sequence because a different initial state is used to reset the shift register in the CM PRN code generator for each satellite. Figure 12.4 depicts the CM PRN code generator with

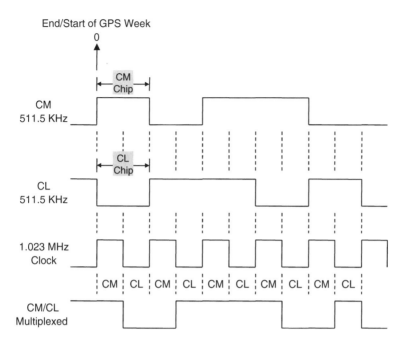

Fig. 12.3 CM/CL multiplexing

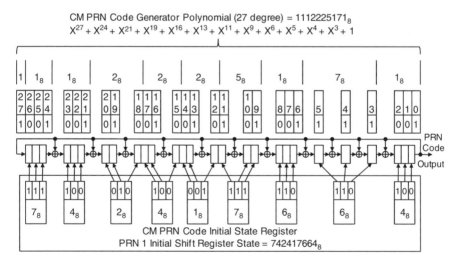

Fig. 12.4 CM PRN code generator

Table 12.6 CM PRN code initial states

PRN	Initial state (octal)	PRN	Initial state (octal)
1	742417664	20	120161274
2	756014035	21	044023533
3	002747144	22	724744327
4	066265724	23	045743577
5	601403471	24	741201660
6	703232733	25	700274134
7	124510070	26	010247261
8	617316361	27	713433445
9	047541621	28	737324162
10	733031046	29	311627434
11	713512145	30	710452007
12	024437606	31	722462133
13	021264003	32	050172213
14	230655351	33	500653703
15	001314400	34	755077436
16	222021506	35	136717361
17	540264026	36	756675453
18	205521705	37	435506112
19	064022144		

Table 12.7 Summary of CM PRN Code

Code chip rate (chips/s)	Code length (chips)	Code period (s)	Chip length (cycles/chip)	Chip length (nsec)	Chip length (m)
511,500	10,230	0.020	2,400 × L2	1,955.03	586.10

the initial state for PRN 1 loaded into the CM PRN code initial state register. The diagram also identifies the position of the exclusive-OR gates as defined by the generator polynomial.

The initial state for the CM PRN code generator initial state register for PRN 1–37 is defined in Table 12.6. Note that PRN 33–37 are reserved and are not assigned to operational satellites.

The first chip of the CM PRN code sequence is synchronized to the start of the GPS week epoch (time of week equal to 0). A summary of the features of the CM PRN code is provided in Table 12.7.

CM Navigation Data

As defined in Table 12.3, the CM PRN ranging code has navigation data modulated on the signal. However, prior to IOC, navigation message data modulation may be disabled (dataless) on the CM PRN code. Legacy navigation (NAV) data with or without Forward Error Correction (FEC) encoding and civil navigation (CNAV)

Table 12.8 Types of navigation data on the CM ranging code

Type of message	Data rate (bits/s)	Symbol rate (symbols/s)	Message size (bits)	Message size (symbols)
NAV without FEC	50	Not Applicable	300	Not Applicable
NAV with FEC	25	50	300	600
CNAV with FEC	25	50	300	600

Table 12.9 Legacy Navigation Data Subframes

Subframe	Description of data
1	GPS week and time and satellite health
2	Ephemeris (precise satellite orbit data)
3	
4	Almanac (coarse satellite orbit and status and error correction data)
5	

data with FEC encoding may be modulated on the CM ranging code. A summary of the types of navigation data messages and transmission rates that may be modulated on the CM ranging code are listed in Table 12.8.

Legacy Navigation Data

Legacy navigation data messages are transmitted at a 50 bits/s rate. Each data bit is 20 ms in length. A NAV data message is composed of 300 bits. Therefore, a message is transmitted every 6 s (300 bits × 0.020 s/bit). Note that the CM PRN code length and the NAV data bit are the same length, 20 ms.

Every 300 bits of a NAV message is referred to as a subframe. Each subframe is divided into ten 30-bit words. Each word contains 24 data bits and 6 parity bits. Subframes are categorized into five types as listed in Table 12.9.

A group of five subframes is referred to as a frame. A frame is 1,500 bits long and 30 s in duration. A group of 25 frames is referred to as a superframe. A superframe is 12.5 min in duration (25 frames × 30 s/frame). The navigation data messages collected during the 12.5-min superframe time interval form a complete set of almanac/ephemeris data for all of the satellites in the GPS constellation

Civil Navigation Data

Civil navigation data messages are transmitted at 25 bits/s, half the rate of legacy navigation messages, due to FEC encoding. FEC encoding converts every CNAV message data bit into two symbols. A bit is a binary digit that represents information. A symbol is a binary digit that represents encoded information. The symbol

Table 12.10 CNAV message types

#	Message type	Description of data
1	0	Default message
2	10	Ephemeris 1
3	11	Ephemeris 2
4	12	Reduced almanac
5	13	Clock differential correction
6	14	Ephemeris differential correction
7	15	Text
8	30	Clock, ionospheric (IONO), and group delay
9	31	Clock and reduced almanac
10	32	Clock and earth orientation parameters (EOP)
11	33	Clock and coordinated universal time (UTC)
12	34	Clock and differential correction
13	35	Clock and GPS/GNSS time offset (GGTO)
14	36	Clock and text
15	37	Clock and Midi almanac

rate is 50 symbols/s because the fundamental navigation data rate is 50 bits or symbols per second.

Because FEC encoding converts every data bit into two symbols, each data bit is 40 ms in length. Each symbol is 20 ms in length. A CNAV data message is composed of 300 bits. However, after it is encoded and transmitted, the CNAV data message is composed of 600 symbols. Therefore, a message is transmitted every 12 s (600 symbols × 0.020 s/symbol). Note that the CM PRN code length and the CNAV data symbol are the same length, 20 ms.

Every 300-bit CNAV message contains 276 information bits and 24 bits of Cyclic Redundancy Check (CRC) parity that covers the entire 300-bit message. The CNAV message structure consists of individual messages which can be broadcast in a flexible manner with a variable repetition sequence. There are 15 types of CNAV data messages that are defined, but the structure permits up to 63 different types of messages. The CNAV messages types that have been defined are listed in Table 12.10.

The use of the 15 different types of CNAV messages will be phased in incrementally leading up to IOC. However, L2C capable satellites in the GPS constellation currently only transmit message type 0, the default message. Message type 0 is a 300-bit message which contains the preamble (a binary pattern, 1000 1011, used for synchronization), the satellite's own PRN number, the message type ID (0), the GPS week number, the GPS time of week, alternating 1 s and 0 s, and the CRC parity block. The GPS week number will increment week to week, the GPS time of week will increment 12 s from message to message, and the CRC will be recomputed for each message.

Table 12.11 Definition of convolutional code parameters

Parameter	Description
k	The number of input bits going into the convolutional encoder
n	The number of output symbols coming out of the convolutional encoder
K	The constraint length represents the length of the convolutional encoder (the number of stages feeding the combinatorial logic that produces the output symbols)

Table 12.12 L2C convolutional encoder parameters

Parameter	Values
K	1
N	2
K	7
Rate	½
Generator Polynomials	$G1 = 171_8, G2 = 133_8$

Forward Error Correction

Forward Error Correction (FEC) is used to encode legacy navigation and civil navigation data messages. The purpose of FEC is to improve the capacity of a channel by adding redundant information to the data being transmitted through the channel. This process increases the probability that the data will be correctly recovered at the receiver end and increases the integrity of the link. The process of adding redundant information to the transmitted data is known as channel coding. One of the major forms of channel coding is convolutional coding (block coding is the other major type). Convolutional codes are used to continuously encode a stream of input data bits. The codes are called "convolutional" because the input data bits are convolved with a generator polynomial to form the output code. Convolutional codes are commonly specified by the parameters listed in Table 12.11.

The ratio of k/n is used to express the code rate. For L2C signals, the convolutional encoder used to encode the NAV and CNAV data messages employ the parameter values listed in Table 12.12.

Convolutional encoding is generated in hardware with a shift register and exclusive-OR gates. The exclusive-OR gates perform modulo-two addition of the inputs and are positioned at locations defined by the polynomials, G1 and G2. Figure 12.5 depicts the L2C convolutional encoder hardware.

Data streams that are encoded with convolutional codes are decoded with the Viterbi decoding algorithm, devised in 1967 by Andrew J. Viterbi.

12.3.2.2   Civil Long Ranging Code

The Civil Long (CL) ranging code is one of the PRN codes available in L2C. The CL PRN sequence is generated as a linear binary pattern that repeats every 1.5 s. The code chip rate is 511,500 Hz. Therefore, the code length is 767,250 chips (511,500 chips/s × 1.5 s).

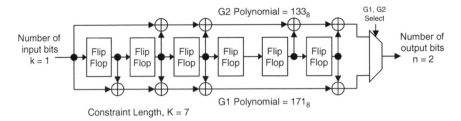

Fig. 12.5 L2C convolutional encoder

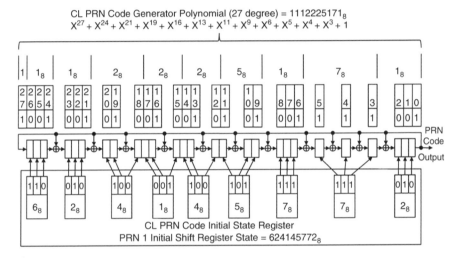

Fig. 12.6 CL PRN code generator

The CL PRN sequence is generated with exactly the same hardware as used in the CM PRN code generator, a shift register and exclusive-OR gates. The exclusive-OR gates are positioned at the same locations as the CM PRN code generator because the same generator polynomial is used. The CL PRN code generator is the same $27°$ generator polynomial, 1112225171_8. The shift register is clocked at a rate of 511,500 chips/s. It is reset every 767,250 chips to an initial value stored in the CL PRN code initial state register. Just like the CM PRN code, a different initial value in the CL PRN code initial state register is used to generate a unique CL PRN code sequence for each satellite, which depicts the CL PRN code generator with the initial state for PRN 1 loaded into the CL PRN code initial state register (Fig. 12.6).

The initial state for the CL PRN code generator initial state register for PRN 1–37 is defined in Table 12.13. Note that PRN 33–37 are reserved and are not assigned to operational satellites.

A summary of the features of the CL PRN code is provided in Table 12.14.

Table 12.13 CL PRN code initial states

PRN	Initial state (Octal)	PRN	Initial state (Octal)
1	624145772	20	266527765
2	506610362	21	006760703
3	220360016	22	501474556
4	710406104	23	743747443
5	001143345	24	615534726
6	053023326	25	763621420
7	652521276	26	720727474
8	206124777	27	700521043
9	015563374	28	222567263
10	561522076	29	132765304
11	023163525	30	746332245
12	117776450	31	102300466
13	606516355	32	255231716
14	003037343	33	437661701
15	046515565	34	717047302
16	671511621	35	222614207
17	605402220	36	561123307
18	002576207	37	240713073
19	525163451		

Table 12.14 Summary of CL PRN code

Code chip rate (chips/s)	Code length (chips)	Code period (s)	Chip length (cycles/chip)	Chip length (ns)	Chip length (m)
511,500	767,250	1.5	$2,400 \times L2$	1,955.03	586.10

Dataless CL

The CL PRN range code does not contain any navigation data modulation. It is referred to as being dataless. A dataless waveform, CL, is easier to acquire than a signal containing navigation data modulation, CM. In this case, CL acts as a pilot carrier which aids in the acquisition of a signal containing data modulation, CM.

12.3.2.3   Coarse Acquisition (C/A) Ranging Code

The Coarse Acquisition (C/A) ranging code is the same PRN code that is modulated on the L1 carrier. The C/A PRN sequence is a Gold code that is generated as a linear binary pattern that repeats every 1 ms. The code chip rate is 1,023,000 Hz. Therefore, the code length is 1,023 chips (1,023,000 chips/s × 0.001 s). The first chip of the C/A PRN code sequence is synchronized to the start off the GPS week epoch (time of week equal to 0). A summary of the features of the C/A PRN code is provided in Table 12.15.

Table 12.15 Summary of C/A PRN code

Code chip rate (chips/s)	Code length (chips)	Code period (s)	Chip length (cycles/chip)	Chip length (ns)	Chip length (m)
1,023,000	1,023	0.001	$1,200 \times L2$	977.52	293.05

Table 12.16 Legacy navigation data framing algorithm without FEC

Step	Description
1	Search for the first preamble (10001011_2). The preamble refers to the first 8 bits of a navigation message. Note that the preamble and its complement should be checked since the data polarity may be ambiguous
2	Verify parity of Word 1 and record the first Time of Week
3	Search for the second preamble (10001011_2) after 300 bits
4	Verify parity of Word 1 and record the second Time of Week
5	If the difference between the first and second Time of Week is one, then the legacy navigation message has been framed

C/A Legacy Navigation (NAV) Data

The legacy navigation data messages that are modulated on the C/A PRN code are the same navigation messages that were previously described in Sect. 12.3.2.1.1.1.

12.4   L2C Navigation Data Demodulation

This section describes the techniques used to demodulate and frame navigation data messages. Navigation data demodulation occurs after PRN code signal acquisition. The data demodulation algorithms that are used depend of the type of PRN code transmitted on L2C. However, the objective of the algorithms is to frame the navigation data messages (find the first bit of a navigation message and continuously collect 300 bits of data).

12.4.1   Legacy Navigation Data Message Framing Without FEC

Legacy navigation data messages are usually modulated on the C/A PRN code. However, it is also permissible to modulate NAV data on CM. The steps required to frame legacy navigation data are listed in Table 12.16.

Table 12.17 Legacy navigation data framing algorithm with FEC

Step	Description
1	Collect a sufficient amount of data to fill the Viterbi trellis (a minimum of 70 symbols)
2	Search for the first preamble (10001011_2) and its complement.
3	Verify parity of Word 1
4	Continue collecting data until the Viterbi trellis is filled again
5	Record the first Time of Week
6	Wait for 600 symbols from the start of the first preamble
7	Collect 70 symbols to fill the Viterbi trellis
8	Search for the second preamble (10001011_2)
9	Verify parity of Word 1
10	Continue collecting data until the Viterbi trellis is filled again
11	Record the second Time of Week
12	If the difference between the first and second Time of Week is two, then the legacy navigation message has been framed

12.4.2   Legacy Navigation Data Message Framing with FEC

Legacy navigation data messages that are FEC encoded are limited to being modulated on the CM PRN code. For this case, a Viterbi decoder, described in Sect. 12.3.4, must be used to decode the navigation data from the satellite. The steps required to frame legacy navigation data with FEC encoding are listed in Table 12.17.

12.4.3   Civil Navigation Data Message Framing

Civil navigation data messages are always FEC encoded and are limited to being modulated on the CM PRN code. For this case, a Viterbi decoder, described in Sect. 12.4.4, must be used to decode the navigation data from the satellite. The steps required to frame civil navigation data with FEC encoding are listed in Table 12.18.

12.4.4   Viterbi Decoder

Forward Error Correction (FEC) is used to encode legacy navigation data messages on C/A and civil navigation data messages on CM. The coding technique is referred to as convolutional coding. Convolutional codes are used to continuously encode a stream of input data bits, such as navigation data, and are decoded with the Viterbi decoding algorithm, devised in 1967 by Andrew. J. Viterbi. The Viterbi algorithm is a maximum-likelihood decoding procedure for convolutional codes.

In the Viterbi decoding algorithm, a convolutional code may be decoded by selecting the path in a time-indexed version of the encoder's state diagram, called a

Table 12.18 Civil navigation data framing algorithm

Step	Description
1	Collect a sufficient amount of data to fill the Viterbi trellis (a minimum of 70 symbols)
2	Search for the first preamble (10001011_2) and its complement
3	Continue collecting data until the Viterbi trellis decodes 600 symbols into 300 bits
4	Verify that the CRC in the message is correct
5	Record the first time of week
6	Collect 70 symbols to fill the Viterbi trellis
7	Search for the second preamble (10001011_2)
8	Continue collecting data until the Viterbi trellis decodes 600 symbols into 300 bits
9	Verify that the CRC in the message is correct
10	Record the second Time of Week
11	If the difference between the first and second Time of Week is two, then the civil navigation message has been framed

trellis diagram Fig. 12.7, whose output bits are closest to the received encoded sequence, yielding the most probable or maximum-likelihood output sequence.

To determine how close the received sequence is to the output of a path, a metric is calculated. A metric is a mathematical function that calculates the distance between two code words. A code word is the symbols that are transmitted in place of the original data. Two types of metrics are used: Euclidean and Hamming distance. Euclidean distance is the straight line separation between two code words, and it is used whenever the received data stream is represented by multiple bits, referred to as a soft decision. Hamming distance, named after R. L. Hamming, is the number of bit positions which are different between two code words, and it is used whenever the received data stream is represented by a single bit, referred to as a hard decision. At each level of the trellis, distance metrics are computed for all paths to a state in the trellis, and the path with the minimum metric is retained. The paths that are retained are called the survivors. After all the metrics and survivors have been calculated for the trellis, the traceback portion of the algorithm is executed. During traceback, the algorithm traverses the trellis backwards to find the predecessor states so that the input bit for each transition may be calculated.

The Viterbi decoder is executed whenever the L2C receiver is acquiring and tracking L2C navigation messages that have been FEC encoded. Since the receiver has no method to determine which type of navigation data, if any, has been modulated on L2C capable satellites, the Viterbi decoder is always enabled during the acquisition phase of satellite tracking. The Viterbi decoding algorithm is partitioned into two segments as listed in Table 12.19.

The Viterbi algorithm operates on symbol (bit) pairs. Once the algorithm receives a symbol pair, it executes the ACS portion of the algorithm where an accumulated error history table and a survivor state history table are calculated. The ACS part of the algorithm must be executed a minimum number of iterations in order to reliably recover the original transmitted binary data stream. The minimum number depends upon the constraint length of the convolutional encoder. The constraint length of the L2C convolutional encoder is 7. In order to reliably recover

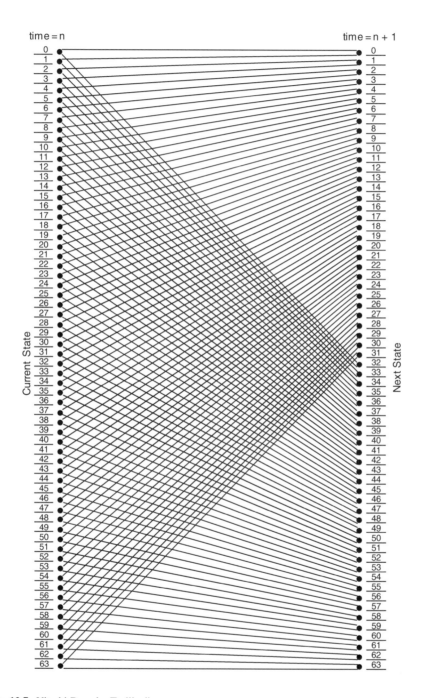

Fig. 12.7 Viterbi Decoder Trellis diagram

Table 12.19 Portions of the Viterbi Algorithm

Parts of Viterbi Algorithm	Description
Add–Compare–Select (ACS)	The Add–Compare–Select (ACS) portion of the Viterbi decoder generates an accumulated error history table and a survivor state history table
Traceback	The traceback and final output portion of the Viterbi decoder generates the final decoded binary output sequence after tracing back through the Viterbi trellis

Table 12.20 Definition of Observables

Observable	Description
Pseudorange	The estimated range (due to the signal propagation delay measured between two non-synchronized clocks) between the GPS satellite and the receiver
Phase	The cumulative sum of the number of cycles and fractional portions of cycles that are accumulated from the start of a satellite track
Doppler	The change in the apparent frequency of the carrier wave due to the relative motion between the GPS satellite and the receiver

the original data, the ACS algorithm must be executed at least five times the constraint length. Therefore, the minimum number of iterations is 35 iterations. Furthermore, the Viterbi trellis must also be 35 states in length. However, the choice of the length of the Viterbi trellis may be longer than 35 states. Extending the Viterbi trellis is useful after acquisition in order to match the length to a multiple of the number of navigation bits in a message. After 70 symbols are processed through the trellis, the traceback portion of the algorithm is executed where the final decoded binary data is generated.

12.5   Observables

This section describes the techniques used by L2C receivers to generate measurements or observables for all of the L2C range codes. The L2C range codes are CM/CL and C/A. The observations are generated for each of the tracked satellites, up to the maximum number of channels in the receiver or the number of satellites that are visible to the receiver. There are three observables that will be discussed in this section and are defined in Table 12.20.

12.5.1   Pseudorange

The technique used to calculate the L2C pseudorange observable in this section will be via direct hardware measurement. The pseudorange value is usually smoothed due to noise and should be referenced to GPS time or time tagged to GPS time. The time tag interval is typically 1 s; every 1 s a pseudorange observable is generated. Atmospheric corrections will not be discussed in this section. The pseudorange is calculated for every channel tracking a visible L2C satellite and for the CM/CL or C/A range codes, depending on the range codes transmitted by the satellite. For this section, the accuracy of the pseudorange measurement is limited by the frequency of the clock used in the hardware.

The hardware generates a measurement every range code epoch. The range code epoch is defined as the length of the PRN code. The epoch of the CM range code is 20 ms; the epoch of the CL range code is 1.5 s; and, the epoch of the C/A range code is 1 ms. When the epoch occurs, the measurement may be read from the hardware and converted to engineering units. The basic pseudorange equation for the CM/CL and the C/A range codes is defined as follows:

$$\text{Pseudorange} = (\text{H/W register - ambiguities}) \frac{c}{f_{\text{CLK}}}$$

where H/W register is the value stored in a hardware register representing the number of clock cycles with respect to a 1 Pulse Per Second (1 PPS) GPS time marker, ambiguities are multiples of the epoch period in counts, c is the speed of light (299,792,458 m/s), and f_{CLK} is the frequency of the clock driving the hardware. For this example, f_{CLK} is 30 MHz.

The hardware that measures the pseudorange is composed of a single 1 s counter and three registers per channel in the receiver. The 1-s counter must be at least 25 bits in length (in order to count 30,000,000 cycles) and is clocked by the 30 MHz signal. The counter is cleared (back to zero) at the leading edge of the 1 PPS pulse and is incremented by 1 count for each cycle of the 30-MHz clock. The registers are separated into sets of three registers (one for CM, CL, and C/A). The 1 PPS pulse is generated by a clock which is synchronized to GPS time. Therefore, the leading edge of the 1 PPS pulse represents the time at which the satellite is transmitting the leading edge of the first bit of a navigation message. Figure 12.8 below depicts the counter and registers in the hardware used to measure the pseudorange.

At every range code epoch, the registers latch the current value of the 1 s counter and store the result for retrieval by a processing unit that calculates the pseudorange. The value that is recorded in the registers represents the number of 30-MHz clock cycles between the leading edge of the 1 PPS pulse and the start of the range code's epoch, a direct measurement of the pseudorange.

Figure 12.9 shows four coordinate frames that depict the relationship between the 1 PPS pulse, the 1 s counter incrementing, and the CM, CL, and C/A epochs. The X axis for all four coordinate frames represents time. The top coordinate frame

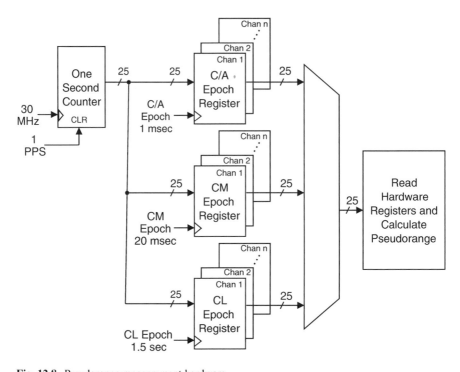

Fig. 12.8 Pseudorange measurement hardware

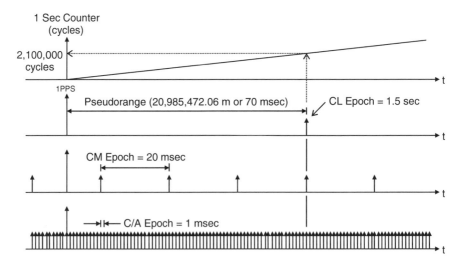

Fig. 12.9 CL, CM, and C/A epochs with respect to 1 PPS

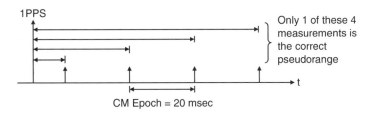

Fig. 12.10 Example of ambiguous pseudoranges

depicts the number of 30-MHz clock cycles (the Y axis) as a function of time (the X axis). When the epoch occurs as shown in the lower three coordinate frames, the number of clock cycles that have transpired between the 1 PPS pulse and the epoch are stored in registers. As an example, the second coordinate frame shows the occurrence of the CL epoch. When the CL epoch occurs, the CL epoch time tag register stores a value of 2,100,000 cycles with respect to 1 PPS. To convert the number of clock cycles to time, divide the number of cycles by 30,000,000 cycles per second, as follows:

$$t = \frac{2,100,000 \text{ cycles}}{30,000,000 \text{ cycles/s}}$$

$$t = 70 \, \text{ms}$$

To obtain the pseudorange from the CL range code register, multiply the measured time delay by the speed of light:

$$\text{pseudorange} = 299,792,458 \, \text{m/s} \times 0.070 \, \text{s}$$

$$\text{pseudorange} = 20,985,472.06 \, \text{m}$$

The pseudorange, however, is not calculated for every epoch of a given range code because it would be ambiguous by multiples of the epoch period. For example, Figure 12.10 depicts four different pseudoranges that would be calculated because the CM epoch occurs every 20 ms.

Pseudorange ambiguity arises because the range code epochs occur at a different rate than the transmission from the satellite to the receiver. Therefore, the pseudorange calculation must compensate for the number of ambiguities present in the hardware measurements. The number of ambiguities may be calculated by searching for the beginning of a navigation message. For the CM epoch, the navigation data bits and the CM range code period are both 20 ms. Once the first epoch is determined for calculating a valid pseudorange (no ambiguities), subsequent CM epoch register readings must have multiples of the epoch (in counts) subtracted to obtain the correct pseudorange as shown in Fig. 12.11.

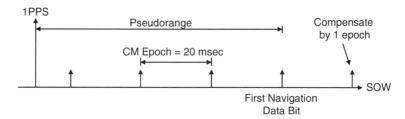

Fig. 12.11 Example of CM epoch ambiguity resolution

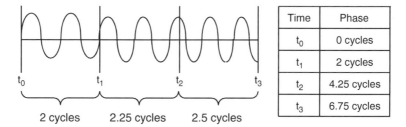

Time	Phase
t_0	0 cycles
t_1	2 cycles
t_2	4.25 cycles
t_3	6.75 cycles

Fig. 12.12 Example of phase accumulation

12.5.2   Phase

The L2C phase observable is calculated by integrating doppler from the beginning of a satellite track. The phase value is usually smoothed due to noise and should be time tagged to GPS time, similar to pseudorange. The phase is calculated for every channel and for each range code (CM/CL or C/A) depending on the range code transmitted by the satellite.

The phase is the cumulative sum of the number of carrier cycles and fractional portions of cycles that are accumulated from the start of a satellite track. An example of the calculation of the phase is depicted in Fig. 12.12. At the start of a satellite track, t_0, the phase is set to zero cycles. At the next sampling interval, t_1, the number of cycles from the initial sample point, t_0, and the next sample point, t_1, are summed (two cycles). At the next sampling interval, t_2, the number of cycles from the prior sample point, t_1, and the next sample point, t_2, are summed (two cycles plus 2.25 cycles equals 4.25 cycles). This accumulation process for the phase continues until the satellite is no longer tracked.

The phase accumulation sequence depicted in Fig. 12.12 may be converted to a mathematical representation suitable for algorithmic computation. To begin, the number of cycles between two successive time instances needs to be computed. The number of cycles is equal to the frequency of the waveform multiplied by the duration of the waveform.

$$\text{Number of cycles} = \text{frequency} \times \text{time interval}$$
$$\varphi = f \times \Delta t$$

Next, the phase at each time interval (ϕ_i) needs to be computed.

$$\varphi_0 = 0$$
$$\varphi_1 = \varphi_0 + f_1 \times \Delta t = f_1 \times \Delta t$$
$$\varphi_2 = \varphi_1 + f_2 \times \Delta t = (f_1 + f_2) \times \Delta t$$
$$\varphi_3 = \varphi_2 + f_3 \times \Delta t = (f_1 + f_2 + f_3) \times \Delta t$$

The phase accumulation may be generalized into the following equation:

$$\varphi_n = \Delta t \sum_{i=1}^{n} f_i$$

The frequency f_i is the signal frequency that remains after the signal has been routed through all of the mixer stages during down conversion. The time interval Δt represents the periodic interval when the signal frequency is sampled.

The frequency f_i that remains after down conversion is composed of the sum of the doppler frequency, f, and a frequency error, f_{Error}. The doppler frequency component of the satellite carrier signal is the frequency induced in the signal due to the relative motion between the satellite and the receiver. The frequency error portion of the carrier signal is the residual phase left after the doppler wipe-off is performed. The frequency error is tracked by a phase locked loop (PLL) for the CL range code and by a Costas loop for the CM and C/A range codes. The PLL phase locks the receiver to the satellite signal. The relationship between f_i, f, and f_{Error} may be expressed as follows:

$$f_i = f + f_{Error}$$

However, the signal at the frequency f_i contains an additional frequency component because it originates in the down conversion process. After the signal is down converted, the signal contains the Intermediate Frequency (IF) component, f_{IF}. Therefore, the carrier signal after down conversion is made up of three components: the IF frequency, the doppler frequency, and the frequency error.

To track this carrier signal, the L2C tracking loop calculates a frequency which is used to down convert the signal down to baseband. The calculated frequency value is then written to a carrier numerically controlled oscillator (NCO), generally performed by a microprocessor. An NCO is an electronic device for synthesizing a range of frequencies from a fixed input frequency source. This device permits the receiver to track the satellite signal by performing the doppler wipe-off and by phase locking the receiver to the satellite signal. The NCO uses a frequency reference as the fixed source, and the frequency resolution is controlled by an

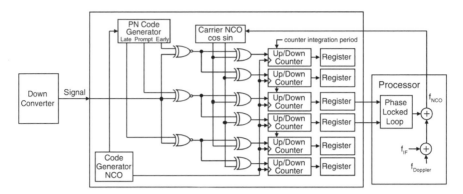

Fig. 12.13 NCO frequency

input from a phase locked loop. Figure 12.13 depicts the combination of the hardware and software required to load the NCO frequency, f_{NCO}.

The f_{NCO} is the frequency value used by the carrier NCO to synthesize a replica sine and cosine carrier signal that is used to down convert the signal from the RF section to a baseband frequency. The f_{NCO} is calculated as follows:

$$f_{NCO} = f_{IF} + f + f_{Error}$$

Combining the formula for f_i and f_{NCO} creates a formula based on the NCO and IF frequencies:

$$f_i = f + f_{Error}$$

$$f_{NCO} = f_{IF} + f_i$$

$$f_i = f_{NCO} - f_{IF}$$

Substituting the equation for f_i into the phase accumulation formula yields the following equation for the phase:

$$\varphi_n = \Delta t \sum_{i=1}^{n} (f_{NCO} - f_{IF})$$

However, the phase is not calculated as expressed in previous equation. If the phase were computed in that manner and converted to distance, the polarity of pseudoranges would be reversed.

To illustrate the polarity issue, an example will be used. If a satellite was being tracked as it was rising, the pseudorange would be decreasing, and the phase would be increasing. Suppose the pseudorange and the phase was measured at two different instances of time. If the pseudorange measured at the second time instance

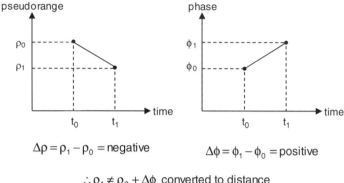

Fig. 12.14 Polarity mismatch between pseudorange and phase

were reconstructed by adding the pseudorange measured at the first time instance and the difference in the phase converted to distance, the resulting value would not match the pseudorange at the second time instance. In fact, the result would be larger instead of smaller in pseudorange. Figure 12.14 illustrates the phase and pseudorange polarity mismatch.

Therefore, to compensate for the reverse polarity between the phase and the pseudorange, the phase formula is multiplied by -1 as follows:

$$\varphi_n = \Delta t \sum_{i=1}^{n} (f_{IF} - f_{NCO})$$

12.5.3 Doppler

The L2C doppler observable is calculated by differencing the IF frequency from the carrier NCO frequency. The doppler value is typically smoothed due to noise and time tagged. The doppler is calculated for every channel and for each range code (CM/CL or C/A) depending on the range code transmitted by the satellite.

As discussed in the phase section, the sum of the doppler frequency, f, and a frequency error, f_{Error}, are the frequency components that remain after down converting the carrier signal. The doppler for the satellite being tracked is reported as the sum of these two frequency components.

$$f_{Reported\ Doppler} = f + f_{Error} = f_{NCO} - f_{IF}$$

The L2C receiver cannot separate the f from the f_{Error} frequency components because the f_{Error} represents errors in the down conversion process. The errors originate from the inaccuracies in the local oscillators during down conversion and are manifest in the resultant doppler values.

Appendix A
Sliding Correlators, Delay-Based Discriminators, and Processing Gain with GPS Applications

This appendix is meant to introduce the reader to the principles of sliding correlators and how they are used to create a delay discriminator. Additionally, the spread spectrum term "processing gain" is discussed. Although there are many other types of correlators, here we will focus on the sliding type. Before we dive right in on the sliding correlator, we need some intro material on the correlation operation.

One of the key operations that distinguish the GPS receiver from classic narrow band receivers is the use of a correlator. The correlation process in the GPS receiver is used to align the replica C/A code with the transmitted C/A code. Additionally, this results in the recovery of timing signals that are ultimately used in the receiver TOA measurement process. But what is correlation and how does it work?

Fundamentally, correlation is a statistical process, that is, it is related to averages and probabilities. We intuitively know that when we roll a pair of dice, the outcome from one throw to another is *not correlated*, that is, the previous throws have no effect on the subsequent throws. When two events are in some way interrelated such that the outcome of one affects the other, we could say that the events have some sort of correlation.

There is another interpretation of correlation and that is as a measure of *similarity*. Particularly in electronics, it would be desirable to compare various time signals to one another and see if they have anything in common. By having such a tool, it should be possible to determine quantitatively *how much and where* two signals are correlated and where *they are not* correlated. It is this interpretation of correlation that is used in the GPS receiver correlation process.

Finding the point in time where two signals are similar, or in GPS receiver where the receiver's replica C/A code is lined up with the received C/A code from the satellite, is the prime reason we need to understand the correlation process. We will see that correlation allows to us determine, to a very high degree of accuracy, when we have C/A code alignment.

Before we discuss correlation in more depth, we need to examine briefly a concept called *"Time Average Value"* of a time varying voltage or current signal.

D. Doberstein, *Fundamentals of GPS Receivers: A Hardware Approach*,
DOI 10.1007/978-1-4614-0409-5, © Springer Science+Business Media, LLC 2012

Time Averaging

What is the "time averaging value" function? Let us look at the waveforms in
Fig. A.1. Figure A.1a shows a sine wave that has a maximum value of + 1 V and a
minimum value of −1 V. Its time average is zero volts. The reason is that the
waveform spends as much time positive as it does negative, and the magnitude of

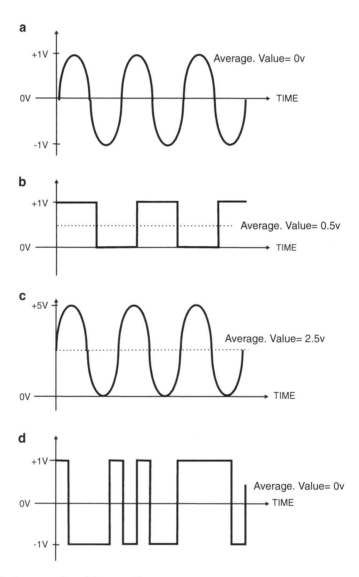

Fig. A.1 Average value of time waveforms

these positive and negative excursions is identical. The waveform shown in Fig. A.1b is a square wave that goes from 0 to 1 V. The average value would be 0.5 V. Waveform A1c is a sine wave that goes from zero volts to 5 V maximum. The average value of this waveform is 2.5 V. Waveform A1d is a small segment of a random binary bit sequence, with values $+1$ or -1. If the sequence is truly "random" and of large length, the average value will be very close to zero. It should be apparent that the time average function of a waveform is the *DC value* of the waveform. If one had a perfect DC voltmeter (reads true independent of waveform type) and applied the waveforms of Fig. A.1 to it, it would read the average value. In lieu of the DC meter, an analog power meter can be used. The total power *minus* the AC coupled power is equivalent to measuring the average or DC value (if properly scaled).

After studying various waveforms and the time average function, the reader should be able to draw a line on the waveform indicating the approximate average or DC value. It is hoped that the reader will develop an intuitive feel for the time average function and be able to approximate this value for most common waveforms.

Correlation, the Mathematical Statement

For the readers who are familiar with convolution, correlation is a closely related operation. Convolution and correlation both use shifting, multiplying, and integration operations on time waveforms (typically). This discussion is restricted to time waveforms only. The mathematical formula for correlation of two time signals is

$$z(t) = \int x(\tau) \times y(t+\tau) \, d\tau \qquad \tau \text{ covers } -\infty \text{ to } +\infty, \qquad (A.1)$$

where $x(t)$ and $y(t)$ are time waveforms for the purposes of this discussion.

The variable τ is the time shift applied to $y(t)$ and the variable of integration.

$z(t)$ is the correlation waveform that we seek.

As usual, writing such equations down does little to inform our intuition on what is really happening! But the math is the exact model that we will attempt to execute in the imperfect world of electronic circuits.

We need to break down (A.1) into its individual operations so that we can better understand what this correlation process does and how we can implement it. First of all, $x(t)$ and $y(t)$ are usually voltage waveforms of the sort you can see with an oscilloscope. If $x(t)$ and $y(t)$ are different waveforms, the correlation of the two is sometimes called *cross correlation*. If they are the same waveform, then the correlation process is a special case called *Auto-correlation*. In the real world of analog electronics, it is almost impossible to have two waveforms to be *exactly* identical, but we can get very close. Auto-correlation is the function we are after for use in the GPS receiver.

Now let us examine the operations needed for (A.1):

(1) Shift $y(t)$ by τ seconds
(2) Multiply $x(\tau)$ and $y(t + \tau)$
(3) Now find the average value of the resulting waveform.

This is the integration process.

So our circuit will need a method to shift a time waveform with respect to another, a way to multiply them together, and a way to time average the result. Of these three operations, the hardest to implement in circuitry is the shifting function. So let us leave that to the end. First, let us look at the multiplying and averaging functions.

A Multiplier and Integrator for Digital Signals

The GPS receiver uses pseudo-random sequences, which are digital signals that are encoded onto the RF carrier. They are digital in that they can only be -1 or $+1$, where here it is better to use a bipolar logic state. We need a multiplier that can take these two digital signals and multiply them together. Such a device is common in the digital logic family; it is just an Exclusive OR gate. We are going to use a modified, bipolar form of the EXOR gate where in/outs are $+1$ or -1. If you let the zero state of the standard EXOR function be -1 instead of zero, you can see that this logic function does a multiplier type function with the output inverted. We chose to ignore the inversion as it is of no consequence and call this model a BI-Polar EXOR (or EX-NOR more properly). The output of the multiplication is $+1$ or -1, as expected. The truth table is shown in Fig. A.2. So now we have our multiplier for our two digital signals; it is just an EX OR gate. The choice of special logic levels makes discussion and modeling easier for using the EXOR function as a multiplier. From a practical point of view, once signals are scaled properly and inversions addressed, the standard EXOR logic gate is equivalent of the defined BI-Polar XOR.

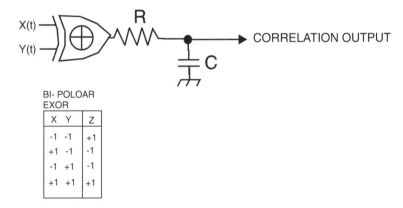

Fig. A.2 A quasi-analog correlator

We can approximate the time average function (or integration) with a simple resistor/capacitor low-pass filter on the output of the EX OR gate. The low-pass filter does not do exact integration, but for our purposes it is close enough. The choice of the time constant of this filter will depend on the code rate, code length, and the rate at which the two signals are sliding by each other. We will come back to this later.

Figure A.2 shows our simple digital correlator. We can supply the two time signals as digital signals to the inputs of the EXOR gate. If we can figure out a way to "time shift" one of them, we will be able to see the correlation process at the output of the low-pass filter. Since our simple circuit uses a digital multiplier with an analog filter, it is neither a pure analog nor a pure digital. The closest description is *quasi-analog*.

Time Shifting or Sliding One Waveform with Respect to Another

Figure A.3 shows two identical C/A code generators. If we use *slightly* different frequencies to clock these two generators, they will appear to slide by each other in time. This is best observed by using a multi-channel scope. Generator #1 is on channel 1 and generator #2 is on channel 2. We trigger the scope on the repeat time of GEN#1 (the C/A epoch pulse). This test setup is shown in Fig. A.3. Assuming that the two clocks for the generators are not equal, the code output from generator 2 will appear to slide right to our left with respect to the code output of GEN# 1 (Chap. 1). *The sign of the frequency difference between the two code clocks will determine the direction of movement, the rate of movement by the magnitude of the difference of the two clocks. If the two frequencies are exactly equal, no movement will be seen but a fixed phase offset of the two C/A codes most likely would be present.*

The reader should understand that if we multiply these two waveforms together point by point and then take the average value, we would only have a nonzero average value when the two codes are within two bits (or chips) of alignment. At all other points, the resulting waveform from the multiplication process has just as many + 1s as –1s so the average is zero. When we have a perfect alignment, the average value would be + 1.

Correlation Pulse and A Delay-Based Discriminator

Figure A.4 shows two identical EXOR-based correlators and two identical C/A code generators. As before, we are offsetting generator #2 code clock slightly so as to slide the C/A code w.r.t. generator #1. Channel one of scope shows the output of correlator #1. As the two codes come within two bits of alignment (two chips), the

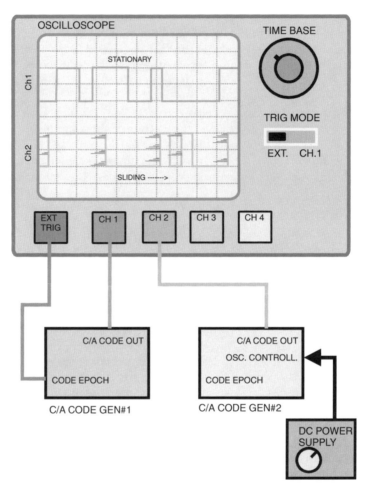

Fig. A.3 Code sliding by difference in code clock FREQ

output starts to rise. As they slide further, a maximum value of +1 is achieved and then we decrease again to zero. Correlator #2 is slightly different in that a delay of one clock time (one chip) has been introduced into C/A code from generator #1. The delay can be implemented in a number of ways, using delay lines, logic gates, clocking, etc. This delay has the effect of moving the correlation peak from correlator #2 to right, or later in time, as shown on channel #2. Channel #3 of the scope shows the difference in waveform obtained by subtracting (point by point) channel #2 from channel #1. This function can be done simply with an OP Amp. Here, we assume that the scope does this function.

The channel #3 waveform is crucial to the GPS receiver's C/A code tracking function. It is a delay-based discriminator (or error voltage) and it forms the heart of the code-tracking loop. If we apply the error voltage properly to the oscillator

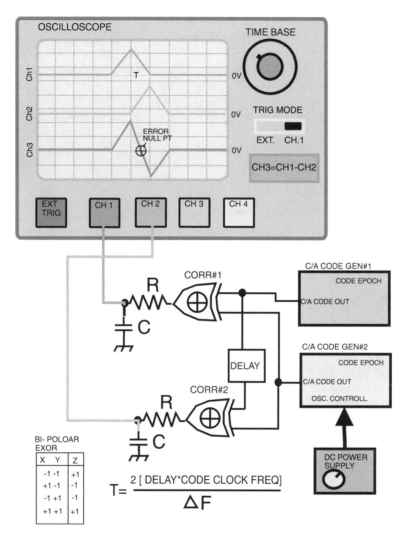

Fig. A.4 Sliding correlators with 1 chip delay discriminator

control point of Gen #2, we can force Gen #2 C/A code to stay aligned with the C/A code from Gen #1. That is, we can lock Gen #2 to Gen #1. This error voltage would obtain code lock at ½ maximum correlation point. This can be remedied by adding another correlator with a delay of ½ chip. This correlator will now be at maximum value when the error voltage channel #3 is used to lock Gen #2 to Gen #1.

The C/A code repeats every 1,023 bits (chips), so the waveforms of Fig. A.4 will repeat in time. The repeat rate of the correlation/discriminator waveforms is

The period of the sliding correlator peaks $= 1,023/\Delta F,$

where ΔF is the frequency difference between code clock Gen #1 and code clock Gen #2. For the C/A code generator, the code clock rate is 1.023 MHz. Typically, ΔF (in analog receivers) is in the 1–10 Hz range, so it is a very small fraction of the code clock frequency. The width of the correlation peak is also determined by the clock frequency difference:

$$\text{Correlation pulse width} = [2 \times (\text{DELAY} \times \text{Code Clock FreQ})]/\Delta F,$$

where DELAY is in seconds and code clock frequency is in hertz. Typically, a one-chip delay is used. A one-chip delay is equal to one code clock period. Therefore, for this case,

$$\text{Code clock period} \times \text{code clock FreQ} = 1$$
$$\text{and correlation pulse width} = 2/\Delta F$$

The Error Voltage or Discriminator Output

The error voltage shown on channel #3 of our scope is shown in more detail in Fig. A.5. Instead of time on the x-axis, the unit of chips is used. One chip is equivalent to 1 bit of C/A code. Therefore, the C/A code has 1,023 individual chips. From Fig. A.5, we see that the error voltage is only linear over a 1 chip transition. Outside this, it reverses sign and returns to 0 V. This lack of range means that a "hunting" technique must be used to get close to lock point and then allow lock to occur, hence the sliding correlator. In practice, the receiver slides its code in

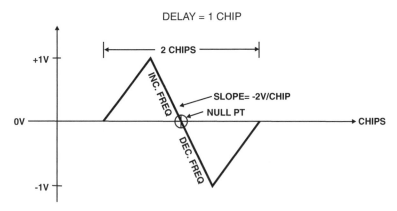

INCREASE/ DEREASE: COMMAND POLARITY MAY INVERT
DEPENDING VCO GAIN AND OTHER INVERSES OF ACTUAL
CIRCUIT..

Fig. A.5 Early late discriminator curve

one direction at a steady state until correlation is detected, i.e., a fixed voltage at Gen #2 controls its oscillator. The receiver then allows the error voltage to control Gen #2 oscillator, thereby initiating C/A code lock.

The linear section of the error waveform can be used to command the frequency of Gen #2 clock oscillator correctly so as to maintain lock. If the error voltage is just *above* null point, the frequency of Gen #2 clock is *increased*. If the error voltage is just *below* null point, the frequency of Gen #2 clock is *decreased*. It may be that an inversion is present and in this case, the above two statements are reversed in polarity of response to error voltage. Note that the discriminator error voltage provides both the *sign* of the error and its *magnitude*.

Similarities to PLL

In many ways, the C/A code correlation/discriminator/PN tracking system is equivalent to a carrier recovery system using a phase locked loop (PLL). A PLL recovery of 1.023 MHz carrier using a 1 kHz comparison frequency is a good model to use for the GPS code clock recovery. In fact, if the equations for the second-order PLL systems are properly re-scaled to reflect the delay discriminator, they can be used to predict the performance of the C/A code track function. The only real difference between the two systems is the phase detector; in the PLL, it is usually a phase/frequency detector. In the PN code tracking systems, it is a delay-based discriminator.

When one examines the quality of the lock obtained on PN-coded waveforms, it is generally noisy or more jittery compared to a PLL approach. This is due partly to the quality of the phase detection scheme; the delay discriminator is a poor performer compared to the phase/frequency detectors used in modern PLL systems. Modern phase/frequency detectors give error information on a cycle-by-cycle basis. A PN code-based discriminator must have good replica code alignment over *many* bit edges in the entire period of the PN code to be effective. Essentially, in a PN code system, we receive *phase error* information over *many cycles* (code clocks in this case) and not on a *cycle-by-cycle* basis as in the PLL system; hence, more jitter in the PN-coded systems for the same SNR.

RC Time Constant

Up till now, we have not mentioned what values should be used for R and C of our simple correlators. These two components form a low-pass filter with a 3-dB roll-off frequency of 1/RC. The choice of this cutoff frequency, or time constant, will depend on the response desire from the system. If the filtered error voltage is used in a closed loop code tracking system, then the time constant choice will be a complex decision based on many considerations. Some of those are loop bandwidth, SNR of signal, and code jitter in lock. Many receivers will use multiple RC time constants [or digital equivalent] that are switched in depending on the conditions in the receiver. Such a system uses the best loop bandwidth for the current condition in the receiver. For example, scanning for code alignment performs better with a wider bandwidth. Once a lock is established, the bandwidth can be narrowed. Generally speaking, wider bandwidths allow for faster lockup times but have more code jitter in the lock condition.

As a minimum RC time constant, we need to integrate over the length of the code used. For the C/A code, this length is 1,023 bits or approximately 1 ms period. Therefore, the RC time constant should be a minimum of 1 ms.

Tau-Dither Discriminator

Figure A.6 shows a modified form of our delayed-based discriminator. Instead of two separate EXOR gates, one gate is used. The delay path is now switched in and out at a rapid rate by the dither clock. This switching of delay in and then out gives rise to the "dither" part of the name. After the EXOR, the filter is now band-pass filter, not a low-pass filter. This band-pass filter is tuned such that its center frequency is that of the dither clock rate. From here, the output of the band pass is amplitude-detected and hard-limited. The hard-limited signal is EXORed with the dither clock. The amplitude of the band-pass filter output is the *magnitude* of the error voltage. The EXOR output of the dither clock and the band-pass output is the *sign* of the error voltage. Both the amplitude detector and the sign EXOR detector are low-pass filtered before they are used. This removes any high frequency components that would be present from the operations of detection, limiting, and EXORing. The scope shows the outputs of the magnitude and sign filters on channels #1 and #2. Channel #3 is the magnitude channel *multiplied* by the sign channel. The result channel #3 shows the same discriminator curve shape that we saw in the early late case.

Here is what is happening. As before, we let the code from Gen# 2 slide by code from Gen #1 by slightly offsetting the Gen #2 code clock. As the code from Gen #2 is sliding by Gen #1 code, it is being dithered back and forth by the delay, typically about ½ chip. This dithering will produce an amplitude modulation waveform on top of a DC level at the output of EXOR correlator. This AM signal will be at the same frequency as the dither clock. The band-pass filter picks out this signal and passes it on to the detector (magnitude) and EXOR/low-pass filter (sign). The AM signal will undergo a 180° phase shift when the two codes pass the exact alignment point. This 180° phase shift contains the error voltage *sign* information and is recovered by EXORing the hard-limited band-pass output with dither clock and subsequent low-pass filtering. The outputs are shown on the scope for the case where the delay is approximately ½ chip. The scope shows the output for the sign portion as zero outside the 2-chip correlation time window.

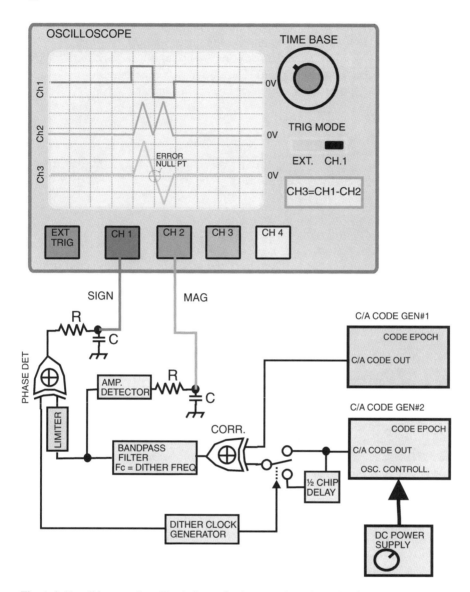

Fig. A.6 Tau-dither correlator/discriminator for the case where delay IS 1/2 chip

This is an idealization. In a real circuit, noise would cause this signal to behave erratically outside the two-chip window, it would go between +1 and −1 randomly.

The advantages of the tau-dither method are that it takes only one correlator to implement code tracking. Also, amplitude imbalances that can be present in the two-correlator method are not a factor. In many PN code tracking loops, it is the sign of the error that is used and the magnitude in information is discarded. In this case, the circuitry is simplified by eliminating the magnitude detector.

The delay- based and tau-dither discriminators described above can both be implemented as carrier-based systems. The advantages of the tau-dither method are more dramatic as carrier-based correlators are significantly more complex than the baseband versions.

Carrier-Based Sliding Correlator

The tau-dither method just described is rarely used in the form outlined above. Rather it is used in the carrier-based system. Both the tau-dither and the early late systems we have just covered are *baseband* correlators as no carrier (or IF) frequency was used. For the receiver described in this text, a carrier-based correlation method is used. In order to use the carrier-based system, modulation of a sine wave with the C/A code is required so that its phase is switched between 0 and 180°. Such a device is the binary phase shift keyed (BPSK) modulator (See Appendix C). The BPSK modulator is the basic building block of the carrier-based sliding correlator. BPSK modulators are reciprocal devices in that they can both modulate and demodulate binary phase-shifted carriers. It is the "undoing" of the phase modulation function of BPSK modulators we seek to employ here (Fig. A.7).

Figure A.8 shows a sliding carrier-based correlator. Once again, we have two C/A code generators with Gen #2 clock rate slightly offset from the clock of Gen #1. The major difference from before is that Gen #1 C/A code is modulating sine wave carrier. This output models the GPS transmitter code modulation of the carrier with no 50 Hz data present. This signal is now fed into another multiplier, where GEN #2 code is applied. The output of this second multiplier is now band-pass filtered with a filter that has a center frequency at the carrier frequency used. When the replica C/A code is within two chips of the reference C/A code, the output of the band pass contains a sine wave. The amplitude of this sine wave grows and then decays as the two codes slide in and out of correlation. This is indicated in Fig. A.10 by the Amp/Freq/Time plot and the carrier in the diamond-shaped envelope.

The principle of operation is very simple. Once the first multiplier modulates the carrier, the second multiplier will completely *remove* the carrier modulation *if* the C/A code of GEN #2 is perfectly lined up with the C/A code of GEN #1. In this condition, the output of the second multiplier is the original sine wave (at the input of first multiplier) along with some artifacts from the second multiplier. The band-pass filter serves roughly the same function as the low-pass filter did in the baseband correlators, as an integrator and to filter out unwanted multiplier artifacts. For those familiar with Fourier transform, recall that multiplication by a sine wave results in a frequency shift in the frequency domain. So the baseband signal spectrums are shifted up by the carrier frequency as well as the low-pass filter operation. A frequency-shifted low-pass filter is just a band-pass filter.

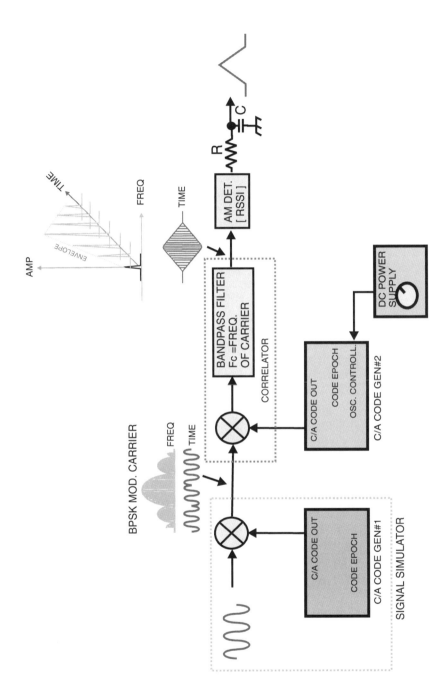

Fig. A.7 Carrier-based sliding correlator

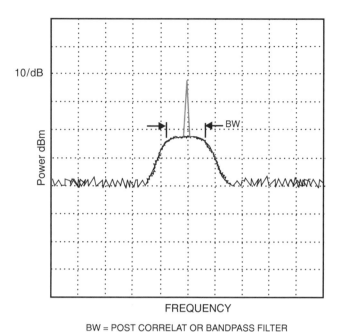

FREQUENCY

BW = POST CORRELAT OR BANDPASS FILTER

Fig. A.8 Post-correlation output of band-pass filter

When the C/A codes from GEN #1 and GEN #2 are not aligned (>two chip offset), the output of the band-pass filter would be the power spectrum of a BPSK modulated carrier. But in a real world receiver, the output of the band-pass filter is noise when the codes are not aligned. The output of the band-pass filter is best examined with a spectrum analyzer, not with a scope as in the baseband case. If a spectrum analyzer is used to look at a carrier-based correlator, the display will typically show a noise pedestal of bandwidth equal to that of the band-pass filter for the case where the codes are not aligned. When the codes are in near or perfect alignment, a spike at the carrier frequency will be seen riding on top of the same noise pedestal. This is shown in Fig. A.9.

A Carrier-Based Tau-Dither Correlator/Discriminator

Figure A.10 shows the Tau-dither correlator using a carrier. The functions of the circuit are very similar to that of the carrier-based system just discussed and of the baseband Tau-dither correlator covered earlier. As in the carrier-based sliding correlator of Fig. A.8, the C/A code is modulated onto the sine wave carrier. The C/A Gen #1 and modulation onto the sine wave are integrated into the first block of Fig. A.10. From here, the signal enters the correlator. As above, the multiplier can be a double balanced mixer (DBM), or many other types as previously discussed.

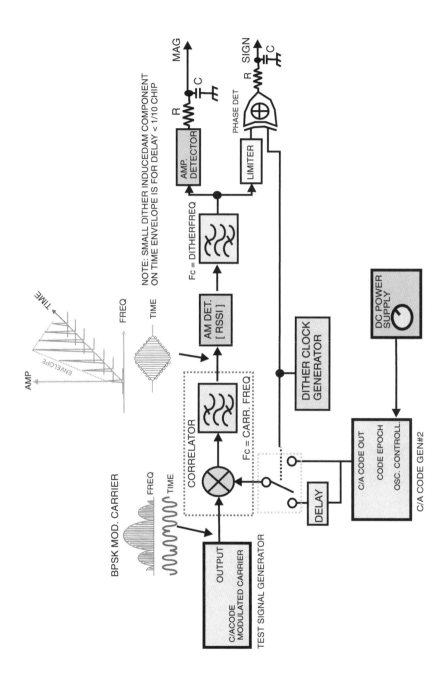

Fig. A.9 Sliding correlator and discriminator, Tau-dither method with carrier

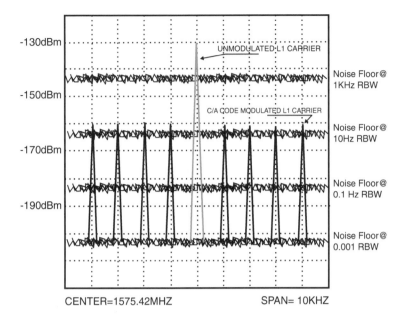

Fig. A.10 GPS carrier with and without C/A code modulation at earth's surface

Gen #2 is offset in frequency by applying the DC control signal to its code clock oscillator. The post-correlation signal is band-passed with a filter whose center frequency is set to the carrier frequency. Typical bandwidths for this filter are in the 200–1 kHz range. The C/A code input to the correlator is dithered between a delayed version and a no-delay version of the code. The dither clock frequency is typically in the 200–300 Hz range for an analog receiver.

Once out of the correlation block, the signal is AM detected (or RSSI function). This signal has an induced dither AM component. The following band-pass filter stage is tuned to the dither clock frequency so as to "pick off" this dither-induced AM signal. In the figure, the time waveform associated with the output of the correlator is a sine wave in a diamond-shaped envelope. This envelope is shown with small AM signal riding on top of the envelope. This would only be valid for small delay values below 1/10 of a chip. Larger value would distort the diamond envelope from that shown in Fig. A.9. Though small values of dither delay can be used, a value between ¼ and ½ chip is more common in GPS receivers.

The output of the dither band pass contains the information needed to create the discriminator function as before. The amplitude of the dither-induced AM has the error magnitude. The phase of the dither-induced AM has the sign of the error. As before, these signals are low-pass filtered before they are used.

Processing Gain

A common term in spread-spectrum system is *processing gain*. Processing gain is realized at the output of band-pass filter of the sliding carrier-based correlator when code alignment is achieved. Here is why: When the receiver replica C/A code is *not* aligned with the transmitted C/A code, the received GPS signal power at the output of the band-pass filter is spread over approximately 2 MHz of bandwidth (center lobe of C/A BPSK spectrum is approximately 2 MHz). When the receiver C/A code *is* aligned with the transmitted code, the signal power at the band-pass output is now squished into approximately 100 Hz of bandwidth (center lobe of 50 Hz random data spectrum). A rule of the thumb is to use the ratio of these two bandwidths as the processing gain. Processing gain is remarkable; it is as if the signal is amplified *without also amplifying the noise* at the same time!

Processing gain = Bandwidth of uncorrelated signal/bandwidth correlated signal

For a GPS receiver, this works out to (in dB)

Processing gain GPS Rec.(dB) = $10 \log[2\,\text{MHz}/100\,\text{Hz}] \rightarrow +43\,\text{dB}$

This number is an estimate of processing gain. A more accurate estimate is the ratio of the SNR before and after correlation. In addition, various imperfections in the correlator may also degrade this gain. Regardless, the processing gain is large in a GPS receiver and enables the negative SNR environment before correlation to be turned into a positive SNR condition after correlation by assuming typical bandwidths before and after.

Recovery of Signal with Negative SNR

The spreading of GPS signal power by the C/A code can lead to the condition where the signal is below the noise floor in one part of the receiver and above it in other sections after processing. When the signal is below the noise floor at a given point in the receiver, it has *a negative* SNR.

For a user at the earth's surface, the received power from the GPS signal is very low. The specified minimum power at the earth's surface is approximately −130 dBm. This power is the unmodulated carrier power in 1 Hz of bandwidth.

When the signal is modulated with the C/A code and 50 Hz data, this power is spread over a larger bandwidth. Once the signal power is spread over 2 MHz, the crest of C/A code spectrum is well below the –130 dBm unmodulated carrier power. Ignoring the 50-Hz data modulation (data mod off), the carrier power is spread over approximately 2 MHz of bandwidth as 1 kHz equal-spaced tones. The power level of these tones in the main lobe is fairly flat at about –30 dB down from the pure carrier level. This puts these lines at approximately –160 dBm.

A typical post-correlation bandwidth for an analog GPS receiver is 1 kHz. For a perfect receiver, this would put the noise floor at approximately –143 dBm. If we do not have correlation (C/A codes not aligned), the 1-kHz tones will be well below the noise floor at this point in the receiver, resulting in a negative SNR condition. Once we have correlation, the signal power is restored to nearly the unmodulated power level, and for 1-kHz bandwidth, we would see an SNR of about +13 dB. So just because a negative SNR condition is present does not mean that the GPS signal is "gone." It is just hiding under the receiver noise floor waiting for the power of the correlation operation to "resurrect" the signal. The exact SNR after correlation will depend on where in the receiver one is talking about and the effective noise bandwidth at that point. In the data demodulator where the data is stripped off the carrier, the noise bandwidth can be quite narrow, and higher SNRs are achieved, typically of 20–30 dB.

The phenomenon of retrieving a negative SNR signal is hard to swallow. It is the authors feeling that this difficulty can be overcome by understanding what noise is and how it interacts with *discrete* signals. Noise from a spectral point of view is a density. It is not discrete. Therefore, you cannot talk about noise levels or powers quantitatively without specifying the bandwidth that is related to the noise being measured (i.e., the circuit bandwidth). Noise power goes up as bandwidth widens. A discrete signal is completely different. Discrete signals, such as pure sine waves, have a power (or level) that is *independent* of the bandwidth of the circuit. Theoretically, if one measures the power of a pure sine wave through a band-pass filter whose center frequency equals the sine-wave frequency, the power measured *is independent* of filter bandwidth. As we just stated, this NOT the case with noise.

As stated above, the power spectrum of the GPS C/A spectrum is made up of discrete spectral lines spaced at the code repeat rate of 1 kHz for the case where we have only C/A code modulation (data = 1 or 0 forever). These lines trace out the $[\sin(x)/x]^2$ spectrum. Therefore, the spectrum of the C/A-modulated signal is not a density but really a collection of discrete signal lines. What would happen if we looked at the received signal (before correlation) with an ideal spectrum analyzer that could use extremely narrow bandwidths?

Figure A.10 shows an ideal spectrum analyzer display of the GPS carrier at 1,575.42 MHz both unmodulated and with C/A code modulation on it (no 50-Hz data mod). The power levels shown reflect those at the earth's surface using a 0 dB-gain antenna. The unmodulated carrier is shown shaded. Note that once C/A code modulates the carrier, the carrier is suppressed and there would be no signal at that position (in an ideal modulator). Our ideal analyzer does not contribute to the received noise power and has some very narrow resolution bandwidths (RBWs).

For each RBW, the theoretical noise floor of the instrument is shown. Note that at a 1-kHz RBW, the C/A tones are well below the noise floor of the instrument. As the RBW is narrowed, the noise floor drops by 10Log RBW, but the discrete C/A modulation tones remain constant in power/amplitude. At an RBW of 0.1 Hz ,we would be able to see the spectral lines of the C/A-modulated carrier *above* the noise floor. So in theory, it is possible to see the GPS signal before correlation.

It is hoped that this example makes it clear that the signal is still there with negative SNR conditions. It is all related to the circuit bandwidth and its effect on noise power. The processing gain of correlation makes it possible to recover the signal and its data by compressing the received signal power into a smaller bandwidth (allowing a reduction in the circuit bandwidth), thereby creating a positive SNR condition.

Appendix B
Pseudo-Random Binary Codes
and the C/A Code Generator

This appendix gives a rough overview of pseudo-random number (PRN) binary sequence generation, their properties, C/A code generation, and the power spectrum of a C/A code-modulated carrier.

What Is a PRN Code Generator?

A PRN code generator is a device that produces a sequence that is a string of binary digits that closely approximate a true random binary sequence. A long record of the results of a coin toss would result in a truly random binary sequence where heads = 1 and tails = 0. We need a circuit that can produce such binary sequences for use in GPS as the C/A code.

A Simple PRN Sequence Generator

The easiest way to make a PRN binary sequence generator is to use a shift register and an exclusive OR gate. Such a circuit is shown in Fig. B.1. Here, a 7-bit shift register is connected to an EXOR gate. The outputs (taps) at Qa and Qg are EX-ORed together and fed back into the input of the shift register. This feedback operation is what generates the PRN sequence. Once the sequence is started, it continues forever repeating the binary sequence at some predetermined length. Each bit of a PRN sequence is called a "*chip*." If different shift register cells (tap points) are connected to the EXOR, different sequences will be generated. Longer registers are needed to generate longer code repeat intervals. The rate of the PRN code, or "*chip rate*," is set by the externally supplied clock. The faster this clock goes, the faster the PRN code goes.

If the tap points are chosen correctly, the PRN sequences (or codes) that are generated are of *maximal* length. If we were using an N bit shift register, the maximal length sequence would be $2^N -1$. For the case where $N = 7$, this works out to a 127-bit long sequence.

Most of all shift register-based code generators must have a way to load them up with at least one binary 1. If they "wake up" with all zeros in the register, the generator will never start the PRN sequence, resulting in an output that is a string of zeros. In our simple example circuit, we do this with a reset line that sets the register to the "all ones" state. Once done, it does not need to be reset again, the code will keep going forever.

If we decode all the register's individual cells into an AND gate, we will generate the *code repeat epoch*. For the generator we have constructed here, the "all 1's" state only occurs one time in the entire 127-bit long sequence. Hence decoding it will produce a pulse that is one clock width wide with a period equal to the clock rate divided by 127 (for this case).

The epoch pulse can be used as a trigger for your scope. The PRN sequences are nearly impossible to trigger the scope reliably at the same spot in the sequence from sweep to sweep. With the epoch fed to external trigger input of your scope, you will always have a stable trigger. If your scope has delayed sweep, you can trigger on the epoch and adjust the delay to inspect the sequence bits.

Gold Codes

Not all PRN sequences are created equal. One can easily generate codes that perform poorly or have other undesirable issues for the application they may be used in. There are "families" of codes that have been optimized for certain performance issues. One of these code families is called the "Gold Codes." The GPS C/A codes are members of the gold code family. Gold codes are maximal length codes. But they also have other desirable properties. When we generate our PRN codes, we would like them to be as close to truly random as possible. They can never be truly "random" as they repeat. A code that is truly "random" would not repeat, ever. But over the length of our C/A code, we very much want it to exhibit the statistical properties that a truly random code would, at least as close as possible. In addition to this, we would want our C/A code to have very low cross-correlation. That is, when we correlate to different codes, the output is very small. Truly random codes would have very low (zero?) cross-correlation output (See appendix A). In addition to these properties, we want the PRN code used in GPS to have a low auto-correlation value outside of the 2-chip wide auto-correlation peak that occurs at code alignment.

Gold codes are optimized for the features just listed. They have good statistical properties and they have low cross-correlation outputs. They also have low auto-correlation output past the 2-chip wide auto-correlation peak. The low cross-correlation property is particularly important in CDMA systems. At the CDMA receiver, we are

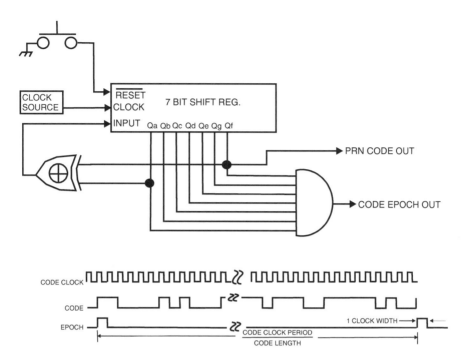

Fig. B.1 Simple PRN code generator with epoch decode

counting on the correlation process to allow us to select which transmitter we wish to listen to and reject the others. As all transmitters are typically on the same frequency (as in GPS), we must reject those we are not listening to. The low cross-correlation of gold codes optimizes the rejection of the other transmitters.

The C/A CODE Generator

The C/A code can be generated by two ten-stage shift registers with appropriate tap points, EXOR logic, and switching, as shown in Fig. B.2. The C/A generator uses two shift registers, each generating a PRN code. These two codes are EXORed together in such a way as to produce the different C/A codes needed. The two tap points selected on register G2 for EXORing with the G1 register output determine a unique C/A code for each satellite.

Tables B.1 and B.2 list the tap points used for each satellite. By changing these tap points, the different C/A codes are generated. If you wish to verify if your C/A code is correct, refer to the last column of Tables B.1 and B.2. It shows the first 10 bits of each C/A sequence. Set your scope to trigger on the epoch (as above) and these bits should be starting at that point in time. In other words, the epoch is aligned with the first bit of the code.

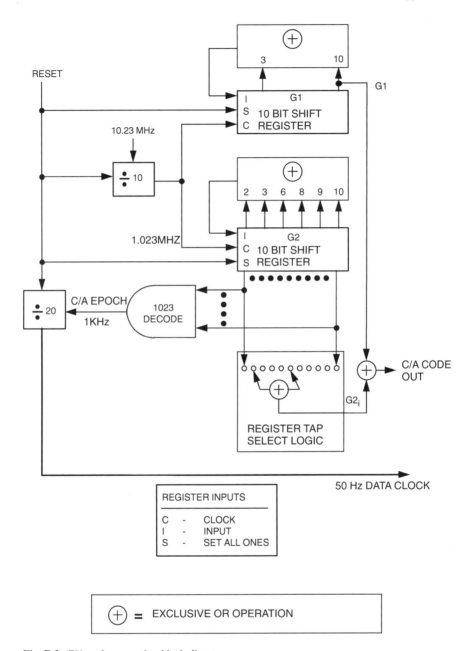

Fig. B.2 C/A code generation block diagram

The reset function indicated is used by GPS to ensure that the generator "starts up" and that it is synchronized with all the other SV generators. The synchronization is very important as it ensures that all SV timing edges (data, code clock, code, etc.) for all SVs are "lined up" at transmission time.

Table B.1 C/A code generator tap points and code samples

SV ID[b] No.	GPS PRN[c] signal No.	Code phase selection C/A (G2$_i$)	First 10 chips octal[a] C/A
1	1	2⊕6	1440
2	2	3⊕7	1620
3	3	4⊕8	1710
4	4	5⊕9	1744
5	5	1⊕9	1133
6	6	2⊕10	1455
7	7	1⊕8	1131
8	8	2⊕9	1454
9	9	3⊕10	1626
10	10	2⊕3	1504
11	11	3⊕4	1642
12	12	5⊕6	1750
13	13	6⊕7	1764
14	14	7⊕8	1772
15	15	8⊕9	1775
16	16	9⊕10	1776
17	17	1⊕4	1156
18	18	2⊕5	1467
19	19	3⊕6	1633

[a] In the octal notation for the first 10 chips of the C/A code as shown in this column, the first digit (1) represents a "1" for the first chip, and the last digits are the conventional octal representations of the remaining 9 chips. (For example, the first 10 chips of the C/A code for PRN Signal Assembly Number 1 are 1100100000)
[b] C/A codes 34 and 37 are common
[c] PRN sequences 33 through 37 are reserved for other uses (Ground Xmitters)
⊕ = Exclusive OR operation

Each SV has a unique C/A code assigned to it. The SVs have an ID number, which is paired with one and only one version of the C/A code. The SV ID numbers are listed in the tables. As mentioned above, the C/A code is a Gold code. It repeats every 1,023 bits as it is based on a 10-bit shift register.

The repeat state is decoded such that when it occurs, a pulse is generated. This pulse is called the C/A epoch. For the C/A code generator shown, the C/A epoch will have repeat a rate of 1 kHz frequency and a pulse width of one 1.023-MHz clock cycle. By dividing the C/A epoch by 20, we generate a 50-Hz signal. This signal is used as the data clock in GPS at both the transmitter and the receiver.

In practice, the GPS receiver would select which SV it wishes to receive. It then selects the tap points from the table and starts generating that C/A code. Hopefully, the SV selected is in view and the receiver can acquire the satellite signal. Once acquired, the receiver can get the data from the SV selected. In this manner, the C/A code acts as an *address* that allows to listen to just one satellite.

Table B.2 C/A code generator tap points and code samples

SV ID No.	GPS PRN signal No.	Code phase selection C/A (G2$_i$)	First 10 chips octal[a] C/A
20	20	4⊕7	1715
21	21	5⊕8	1746
22	22	6⊕9	1763
23	23	1⊕3	1063
24	24	4⊕6	1706
25	25	5⊕7	1743
26	26	6⊕8	1761
27	27	7⊕9	1770
28	28	8⊕10	1774
29	29	1⊕6	1127
30	30	2⊕7	1453
31	31	3⊕8	1625
32	32	4⊕9	1712
c	33	5⊕10	1745
c	34[b]	4⊕10	1713
c	35	1⊕7	1134
c	36	2⊕8	1456
c	37[b]	4⊕10	1713

[a] In the octal notation for the first 10 chips of the C/A code as shown in this column, the first digit (1) represents a "1" for the first chip, and the last digits are the conventional octal representations of the remaining 9 chips. (For example, the first 10 chips of the C/A code for PRN Signal Assembly Number 1 are 1100100000)

[b] C/A codes 34 and 37 are common

[c] PRN sequences 33 through 37 are reserved for other uses (Ground Xmitters)

⊕ = Exclusive OR operation

Power Spectrum of a Carrier Modulated with C/A Code

PRN codes have a complex power spectrum. The envelope of a PRN sequence baseband spectrum is approximately defined by the $[\sin x/x]^2$ function. Inside the spectrum, it is not continuous in nature. If the spectrum is displayed on a spectrum analyzer, it is easy to make the false interpretation that the power spectrum is continuous and that there are no discrete elements to the spectrum. But a closer look would reveal that it is discrete. Inside that $\sim[\sin x/x]^2$ envelope are individual spectral "lines" or tones separated in frequency by the code repeat rate. The individual lines will vary slightly from the $\sin x/x$ envelope due to nonrandom characteristics of the specific PRN code. For the C/A code used in L1 GPS, these variations can be as much as 6 dB.

Figure B.3 shows the C/A code spectrum modulated onto a carrier. This is just the baseband spectrum shifted up to the carrier frequency. We chose to show the spectrum as carrier based. This is typically how it would be observed inside a GPS receiver. The magnified view shows the individual lines separated by the 1-kHz C/A

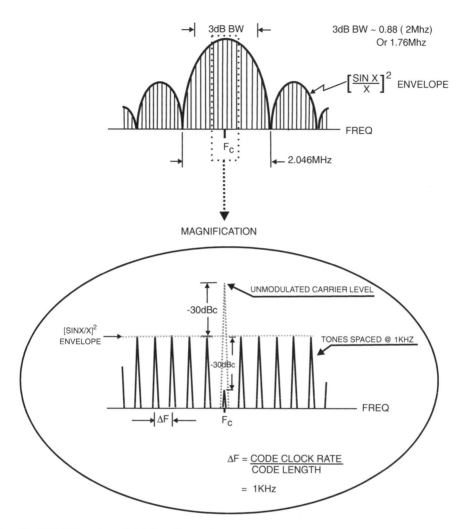

Fig. B.3 C/A code-modulated carrier spectrum

code repeat rate. If we were to lengthen the C/A code, the lines get closer and closer together. For a very long code, it would take a special equipment to see the discrete lines, but they would be there.

Getting the relative power levels of 1-kHz tones and the carrier when we modulate it with the C/A PRN code is a little tricky. Specifically, if we take our carrier and measure its unmodulated power, what will the relative power level be of the 1-kHz tones when we modulate the carrier with the C/A code? Furthermore, what happens to the carrier power?

Let us tackle the carrier power first. If the C/A code were infinite and our modulator perfect, the carrier power would be zero. For a perfect modulator and our 1,023-bit C/A code, the carrier would be down by approximately 1/1,023 factor, or -30 dBc with respect to the 1-kHz tones at the center of the spectrum. This is near the carrier suppression level of many RF-type BPSK modulators (see Appendix C). So we can expect the carrier to be down by no more than 30 dB typically.

We can predict the relative power of the 1-kHz discrete lines by estimating how many of them are needed to absorb most of the unmodulated carrier power. The 3-dB bandwidth of the C/A code-modulated spectrum of Fig. B.3 is approximately 0.88 (2 MHz) = 1.76 MHz. If we assume that this bandwidth now contains most of our carrier power, we can estimate the individual 1-kHz tone relative power near the center of the spectrum. Inside the center lobe of the spectrum, it takes 1,760, 1-kHz spaced lines to occupy 1.76 MHz of bandwidth. Therefore, the 1-kHz tones will be below the unmodulated carrier level by

Relative power of center 1 kHz Tones (dBc) $\approx$ Unmodulated carrier power (dBm)

$\quad - \log_{10}(1/1,760)$

$\log_{10\neg}(1/1,760) = -32$ dB

Summarizing, if we look at the C/A code-modulated carrier spectrum, we could expect to see the carrier down by approximately 30 dB w.r.t. the 1-kHz tones at the center of the spectrum. The center of spectrum 1-kHz tones will be down approximately 30 dB from the unmodulated carrier level. These power levels are indicated in Fig. B.3.

This completes our brief look into the PRN and C/A code generation and properties. The reader should consult other references if more information is needed.

Appendix C
BPSK Modulators and Demodulators

BPSK Modulation/Demodulation Fundamentals

This appendix is devoted to the operation, properties, and attributes of BPSK modulators and demodulators. Figure C.1 shows the basic modulator. The operation is quite simple. The binary waveform applied to the modulation input switches the sine wave (carrier) between 0° and 180°. If the modulation input is a PN sequence, then the resulting spectrum is as shown in Fig. B.1.

All BPSK modulators are reciprocal devices in that they can be used to demodulate a carrier. This is shown in Fig. C.2. Here, a sine wave that is BPSK modulated with a PN sequence is applied to the demodulator, resulting in all the phase transitions being undone. The result is a carrier with no phase transitions, or in this case, a *recovered carrier* is the output. This is actually the fundamental principle behind removing the C/A code on the GPS carrier. Once removed, the data modulation is left for the receiver to decode.

Figure C.3 shows the inside of a BPSK modulator in functional blocks. We see that the carrier is applied to a SPDT switch. One leg of the switch has a phase shift of 180°. The binary modulation input causes this switch to toggle so that either the 0 or the 180° carrier is selected to the output. There are many ways to implement the two basic functions of 180° phase shifting and switching.

Figure C.4 shows three different implementations of BPSK modulators. The modulator shown in Fig. C.4a is a DBM. It uses diodes to do the switching. The RF transforms provide the 0° and 180° versions of the carrier. When a positive current is supplied to the I port of the mixer, one phase is selected. When a negative current is applied to the I port, the inverted phase is selected. The RF port now has the BPSK-modulated carrier on it. A TTL compatible driver circuit is also shown. Its just two 74HCT gates. The second gate is biased with bipolar supply. This allows the gate to drive the positive and negative current into the I port. The series resistor is chosen such that the current into the I port is optimized. The series resistor also provides some isolation. For a 50 Ω I port, this resistor is small, typically 10–20 Ω.

The modulator shown in Fig. C.4b uses a RF SPDT switch and an RF transformer. The RF transformer provides two versions of the carrier input, 0 and 180°. The RF switch switches between these two carrier versions, thus providing a BPSK-modulated carrier at the SPDT throw terminal. The RF transformer can be a wire-wound one or a transmission line type for microwave frequencies.

The last BPSK modulator shown in Fig. C.4c is an EXOR gate. If one input is connected to the carrier (in this case, a digital clock signal) and the other input to the modulating digital source, a BPSK signal will be present on the EXOR output. This type of modulator is for lower IF frequencies and typically cannot perform at RF frequencies.

BPSK modulators have a number of design problems and measures of modulator performance. We will now discuss the most import of these.

Phase and Amplitude Imbalance

In a perfect BPSK modulator, the phase difference is exactly 180°, and the two legs have equal power. In real modulators, this is never achieved. If the phase difference between the two carrier versions internal to the modulator are not 180°, degradation of the modulated waveform will occur. To a lesser extent, the same is true for the amplitude of the two carrier versions. In general, the higher the carrier frequency, that the modulator is operating on, the more difficult it is to assure an adequate phase and amplitude matching.

Isolation

In a perfect modulator, if no carrier were present on the carrier input port, no carrier would be present on the output port of the modulation port. If a carrier is applied to the output port, it should not appear on the carrier input port or the modulation input port. This is reverse isolation. Also, if data are applied to the modulation input port, this signal should not be present on the other ports. In real modulators, signal leakage from lack of isolation from one port to another can cause real problems.

Code Clock Feed-Thru

When BPSK modulators are used to impress a PN sequence like the C/A code onto a carrier, the C/A code clock can leak through to the output port. An easy test of this is to turn off the carrier to the modulator and apply the C/A or other PN code to the modulator. With no carrier, inspection of the output with a spectrum analyzer will reveal if any leakage is occurring. The closer to The carrier frequency is the code

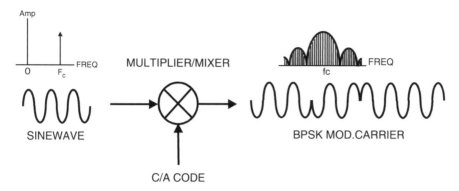

Fig. C.1 BPSK modulator

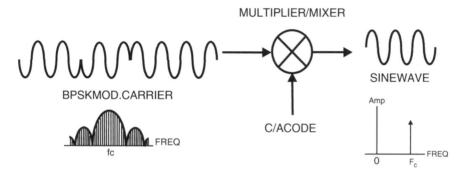

Fig. C.2 BPSK demodulator

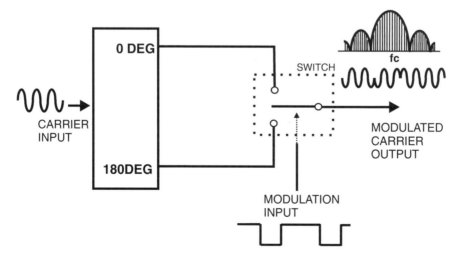

Fig. C.3 Basic BPSK modulator block diagram

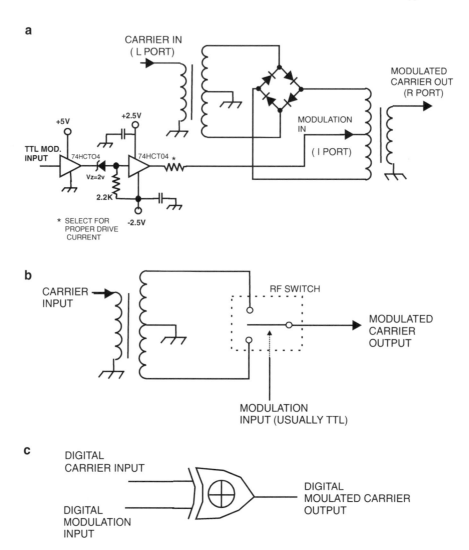

Fig. C.4 (**a**) Double balanced mixer as BPSK modulator with bipolar TTL driver. (**b**) RF switch/ transformer BPSK modulator. (**c**) EXOR as BPSK modulator

clock frequency, the worse this problem can become. This is because of the broadband nature of the PN sequence signal, like the C/A code. This problem is really a manifestation of the isolation properties of the modulator. Figure C.5 shows a test setup for measuring code clock feed-thru.

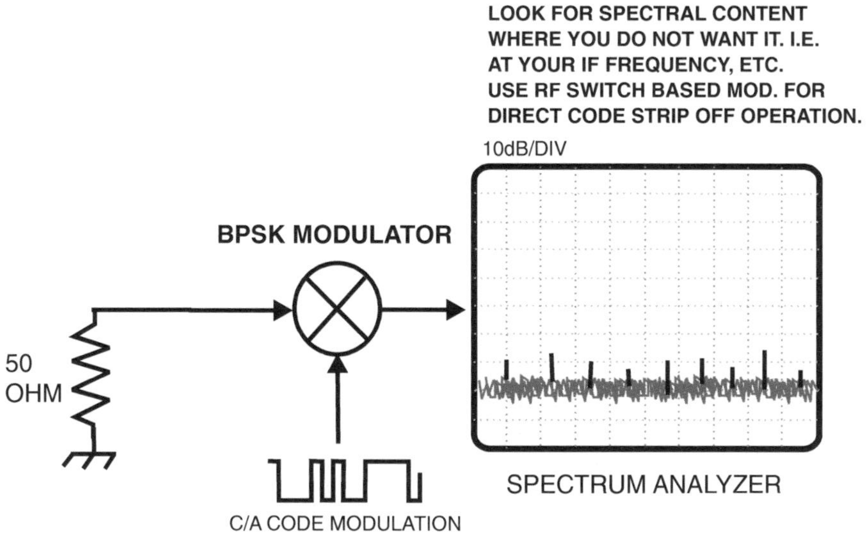

LOOK FOR SPECTRAL CONTENT
WHERE YOU DO NOT WANT IT. I.E.
AT YOUR IF FREQUENCY, ETC.
USE RF SWITCH BASED MOD. FOR
DIRECT CODE STRIP OFF OPERATION.

10dB/DIV

BPSK MODULATOR

50 OHM

C/A CODE MODULATION

SPECTRUM ANALYZER

Fig. C.5 Code feed-thru test setup

Carrier Suppression

Carrier suppression is a figure of merit for a BPSK modulator. Its easy to set up. Apply a square wave to the modulation input port and carrier. Now look at the output with a spectrum analyzer. If the square wave is truly a square and the modulator perfect, there would be no carrier present. All the carrier power is now in sidebands symmetrically about the carrier frequency. Usually, getting a good square wave is not hard, so any carrier that is present is due to imbalances in the modulator. This gets back to the phase and amplitude imbalance issue above. Measuring carrier suppression is a way to check for this. A carrier suppression of 30 dB is considered pretty good. Above 30 dB is exceptional. If you see lousy carrier suppression, it can also be caused by lack of isolation between carrier input and modulator output ports. In other words, if your isolation is limited to 25 dB, good luck trying to get carrier suppression of 35 dB. The carrier leakage will be dominant, and not phase and amplitude imbalance.

Carrier suppression also has another important interpretation for CDMA systems like the GPS receiver. It is called the jamming ratio. When a unwanted sine wave hits the correlator portion of a GPS receiver (i.e., a BPSK demodulator), the sine wave is spread out by the C/A code. In theory, this should reduce the offending signal to very low levels as C/A code modulation suppresses the carrier. In practice, the carrier suppression ability of the modulator may not be good enough to obtain theoretical values of carrier suppression. Figure C.6 shows the test setup for measuring carrier suppression.

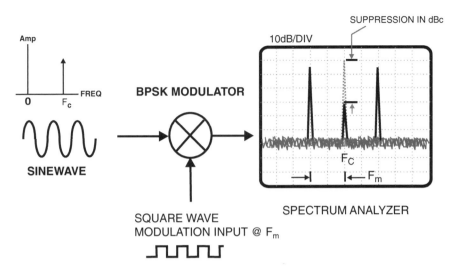

Fig. C.6 Carrier suppression test for BPSK modulator

Relative Merits of the BPSK Modulators

Each of the three modulators shown if Fig. C.4 will be discussed as to the ease of use and performance issues.

Double Balanced Mixer

This is the most common type used in RF and IF circuits. The I port is driven with the modulating signal and the L port with the carrier. The RF port is the modulation output. In practice, the RF and LO ports can usually be switched. The main draw back to using a DBM is that it needs currents at relatively high drive levels to switch the diodes on and off. This is usually accomplished with a bipolar supply on the driver, i.e., plus and minus voltages. Logic gates can be used (74HC series works well), but they must be biased below ground potential to allow for bipolar current flow.

The drive issue can be solved easily in the case where a high rate signal is applied to the modulation port. In this case, a 74HC gate is AC coupled to the modulation port. As long as the AC coupling can pass the lowest frequencies contained in the modulation signal, this will work. If, on the contrary, the modulation signal has very long periods of all 1s or all zeros, this solution has a problem. When a long series of 1s or 0s hits the AC coupling, they do not pass. So AC coupling can work but it has limitations.

Another problem with DBM-based BPSK modulators is poor isolation. The best they can usually do is around 30 dB. This can create code clock feed-thru issues as well as other problems in the IF or RF signal processing.

A final limitation of DBM modulators for BPSK is that of power handling. At certain carrier input power, they start limiting or saturating. This translates to a maximum power they can pass to around 0 dBm. With a typical loss of 3–5 dB in the modulation process, the carrier at the input cannot exceed approximately +7 dBm. The lower power levels may mean that additional amplifiers will be needed to increase the signal level subsequent to modulation.

Transformer (Or Balun)/Switch

This is a newer method made possible by readily available GASFET and other types of integrated RF switching devices. These switches, available as SPDT devices, are fast, have high isolation, and have higher carrier power handling than a DBM. In particular, the isolation from the modulation port to the carrier input port and the modulation output port can be extremely high. This is due to the low current levels needed to switch many of the devices. Lower current levels mean that higher series impedance is possible by increasing the isolation. These lower currents translate to lower power consumption. The drive current required can be in the microamps compared to 20–30 mA for the DBM. As a final plus, many of the switches available have inputs that accept TTL level signals, so no level shifting of logic level is needed.

On the down side, this method is not as integrated as the DBM (at this time). Another problem is locating small RF transformers or Baluns for the phase shifting function. This can limit the use of this method to the lower RF and IF areas compared to DBM, which can be used well into the GHZ region.

EXOR Gate

The EXOP gate BPSK modulator is very limited. The function of switching the carrier between inverted and non-inverted state is integral to the EXOR truth table, inversion of logic being roughly equivalent to $180°$ phase shift. If the carrier is not applied as a square wave, degradation in the modulation will occur. Reverse isolation from mod output to carrier and mod input is very good. Inherently, the EXOR is limited in its applications to low frequency to low end of RF frequency range depending on what performance is needed.

Appendix D
Subframe Format

Figures D.1–D.4

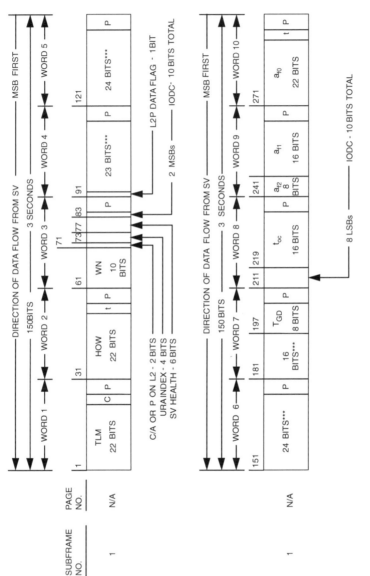

Fig. D.1 Subframe #1 data format

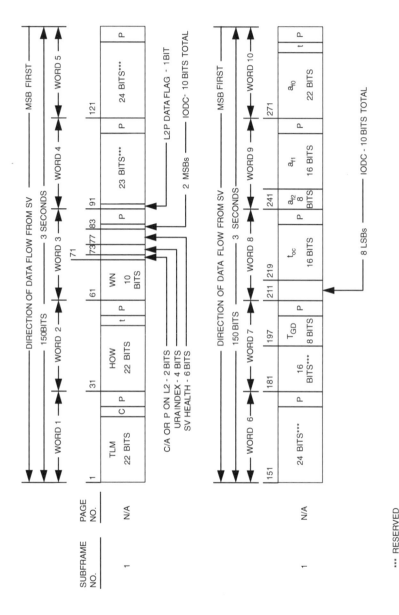

Fig. D.2 Subframe #2 data format

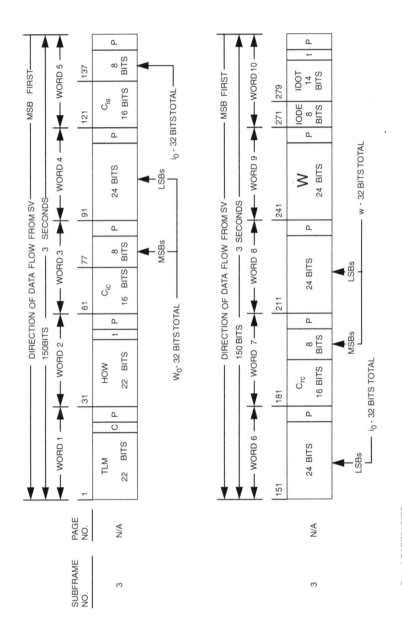

Fig. D.3 Subframe #3 data format

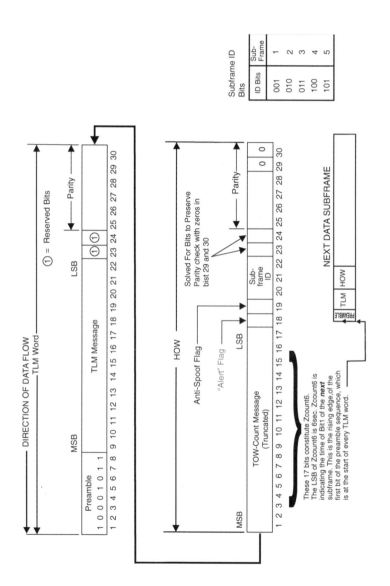

Note: The two 30 bit TLM and HOW are sent at the beginning of each 300 bit long subframe.

Fig. D.4 TLM and HOW formats

Appendix E
Glossary

Carl Carter

Aided GPS, AGPS GPS in an environment where external data sources are used in place of relying only on information from GPS satellites. For example, in a GPS receiver associated with a cell phone, the cell phone network could be used to provide information such as the ephemeris of visible satellites. This would permit the receiver to achieve a fix without having to collect the ephemeris from the satellites and could result in faster time to first fix, or fixes from much weaker signals which might have unreliable data decoding. Typical data supplied in such an arrangement include ephemerides, almanacs, approximate position, approximate or precise time, frequency information for the receiver's crystal, approximate pseudo-ranges, satellite Dopplers, and visibility information.

Almanac A collection of a reduced version of the ephemeris, condensed from 3 to 1 subframe, and broadcast by all satellites over a 12.5-min period. Each satellite broadcasts the same almanac, which covers all satellites. Almanacs are used by receivers to predict satellite visibility.

Ambiguity resolution The process of resolving the integer number of carrier cycles between a satellite and a receiver. This generally cannot be done without external information, such as information from a second receiver monitoring the same satellite at the same time.

Anti-spoofing (A-S) Spoofing is a means of confusing a GPS receiver by sending a false signal from a transmitter masquerading as a GPS satellite. Anti-spoofing, or A-S, is a countermeasure to spoofing. In GPS, it is implemented by encrypting the data stream on the P code signal. The encrypted data stream is referred to as Y code.

Azimuth The angle in the horizontal plane from an observer between a reference direction (usually true North) and an object such as a satellite.

Baseline A line connecting two survey stations. Survey system error is frequently defined as some specific amount plus an amount referenced to the length of the baseline, as in 2 mm plus 5 ppm of baseline length.

Broadcast ephemeris The description of a satellite's orbit and clock corrections as transmitted by that satellite. The information is contained in the first three subframes of each satellite's navigation message. It is composed of one subframe describing the satellite's clock with respect to GPS time, and two subframes defining the orbit. Each satellite broadcasts only its own ephemeris.

C/A code, Coarse and Acquisition code The dominant signal modulated onto the L1 carrier of each satellite. The code is composed of a 1,023-bit long sequence of 1s and 0s in a seemingly random pattern; thus, it is often called a PRN sequence, or a PRN code. The signal is modulated onto the carrier by $180°$ phase shifts whenever the PRN sequence changes between 0 and 1, or 1 and 0. The bits have a data rate of 1.023 MHz (called the chipping rate), so the 1,023-bit pattern requires 1 ms to transmit, and each bit has an effective wavelength of approximately 293 m. Each satellite has its own unique PRN code, called a Gold code. To acquire the satellite signal, receivers must make a copy of the code and time align it with the received signal. Sometimes called the clear and acquisition code since it is never encrypted. See P Code.

Carrier phase The measurement of pseudo-range made by tracking the number of cycles and fractional cycles of the carrier that pass between measurement times. In a stand-alone receiver, this is simply a relative measurement since the total number of integer cycles between the satellite and the receiver cannot be determined. However, the change in distance, that is, the number of cycles that are observed versus the number that should occur based on the carrier frequency, represents the change in range between the satellite and the receiver. This change in range is due to both satellite motion and receiver motion. Since the tracking of the carrier is much less prone to noise, and generally much more precise due to the 0.19-m wavelength, than the tracking of the PRN code ("code phase"), the measurement can be used to smooth the measurements made using the code phase. In receivers used in either post-processed or paired environments, relative measurements between multiple receivers can be used to solve for the total number of integer cycles as well (a process called ambiguity resolution), which will result in significantly more precise ranges.

Carrier wave The radio signal sent by a transmitter. A transmitter generally has a specific frequency at which it transmits, and this frequency is the number of transitions the signal makes per second. The resulting signal travels from the transmitter at the speed of light as a series of alternating fields (electric and magnetic). The fact that the changes occur at a specific frequency and that the changing signal travels at the speed of light (about 3×10^8 m/s) means that the distance between successive peaks of the signal can be calculated. This distance, called the wavelength, is an important characteristic of the signal. For GPS satellites, the L1 transmitter has a frequency of 1,575.42 MHz (million cycles per second), and thus has a wavelength of approximately 0.1903 m in free space. While the speed of light is slightly different in the atmosphere, most of the pathways between the satellite and a ground-based receiver are through free space.

Chip, chipping rate The PRN code that modulates a spread-spectrum signal is used to select a phase of the carrier wave to transmit at a particular time. Each bit of the PRN code sequence is called a chip, and the rate at which the bit stream modulates the carrier is called the chipping rate. Since the phase changes have a specific time of occurrence, it follows that the transitions, when traveling through space from the transmitter, have an associated wavelength. The C/A code used in GPS has an associated chipping rate of 1.023 MHz, and therefore, an associated wavelength of approximately 293 m. The P code has a chipping rate of 10.23 MHz, and an associated wavelength of 29.3 m.

Clock bias, clock offset The time difference between the receiver's reference clock and GPS time. When the receiver measures pseudo-ranges to the satellites, it assumes a clock starting time and measures the ranges as offsets from that assumed starting time. When the navigation solution is computed, the difference between the assumed starting time and actual GPS time is determined just as the receiver's x, y, and z position is computed. Thus, clock bias is considered the fourth unknown in the navigation solution and is the reason a pure GPS solution requires four satellites

Coherency The characteristic of two or more signals that share a common phase relationship over time. If a single reference oscillator is used to generate two or more signals, each of which becomes phase locked to the reference, then the resulting signals may be assumed to be coherent. The L1 and L2 carriers, and the C/A and P codes in a GPS satellite are all generated from the same reference oscillator and are thus coherent. In contrast, signals from two different GPS satellites, while very close in frequency, are generally not coherent since their oscillators are not phase locked to each other, and each will generally have a different Doppler.

Control Segment The network of ground stations, operated by the US Air Force, that monitors the performance of the GPS satellites (the Space Segment) and sends them control signals to maintain the system operation within specifications. The 50th Space Wing at Schriever AFB, Colorado, operates the main control facility, with other sites located around the world so that the entire constellation of GPS satellites is under constant observation.

Cycle slip When tracking the carrier phase of a GPS signal, if phase lock is lost during a specific observation interval, the resulting observation may show a discontinuity compared with observations made before and after. This disconti-nuity is referred to as a cycle slip. Its primary implication is that any ambiguity resolution in progress will need to be restarted.

Data link A connection between two or more devices that allows the transfer of information between the devices. Examples include serial communication circuits between two computers, parallel links as between a computer and a printer, and network connections between multiple computers, printers, and other devices.

Data word In the context of the GPS system, this refers to a grouping of 30 bits of information in the navigation message broadcast by each satellite. Each data word consists of 24 bits which carry information, and 6 bits which permit verification that the information bits were received properly. The data words are organized in groups of ten to form subframes.

DataBit checking A process of testing navigation message data for reasonableness before using it. The process generally involves testing key values against limits that would not be exceeded in a working system, such as an orbital radius that is too close to the Earth, or too far away.

Datum In geodesy and surveying, the specific mathematical model used as the basis for a map or other representation of position. It consists of a reference ellipsoid, and optionally adjustments to that reference ellipsoid such as origin translations which adjust the ellipsoid to more appropriately match the local region of interest. GPS systems use the WGS-84 datum (q.v.), but many receivers have features that permit them to relate position in any of several hundred other defined datums.

Differential GPS, DGPS A technique to improve accuracy in a GPS receiver by using corrections supplied from another receiver called a reference receiver. The reference receiver generally knows its position accurately and is thus able to estimate the error on each satellite measurement. The estimated error is communicated to the using receiver, which uses it to adjust its own measurements before computing a position. The error estimate contains a mix of errors from sources common to both receivers (such as errors in the satellites' clocks or orbit parameters), sources unique to the reference receiver (errors in the reference receiver clock, or due to local multipath at the reference receiver), and errors that have varying amounts of commonality between the two receivers (such as ionospheric errors, which are less common when the distance between the two receivers is larger). Generally, use of such corrections yields improvements that may range from a meter or two under very good conditions, to several tens of meters when selective availability is active in the satellites.

Dilution of Precision (DOP) The effect of geometry and number of satellites on error in a GPS solution. When a position is computed based on the information from a large number of satellites evenly dispersed about the sky, the error is generally smaller than when the solution is based on very few satellites clustered in one part of the sky. DOP is a mathematical estimate of this effect. There are several forms of DOP, including Geometric DOP (GDOP), Position DOP (PDOP), Time DOP (TDOP), Horizontal DOP (HDOP, and Vertical DOP (VDOP). All are interrelated. In use, the error in a particular part of the solution can be estimated by multiplying the system error by the appropriate DOP. Smaller values of DOP imply less error in the computed position or time.

Double difference solution A method of processing GPS observations to eliminate errors. Observations of two different satellites by two different receivers are combined by computing single differences (q.v.) and then taking the difference between those answers. The resulting double difference is free of clock errors in the two satellites and in the two receivers.

ECEF coordinates Earth-centered, Earth-fixed (ECEF) coordinates. A coordinate system which is tied to and rotates with the Earth. GPS solutions are computed in such a coordinate system which has its origin at the mass center of the Earth and the Z axis aligned with the Earth's spin axis. The $+X$ axis passes through $0°$ latitude, $0°$ longitude, and the $+Y$ axis passes through $0°$ latitude, $90°E$ longitude.

Elevation angle The angle between an object and the horizon from an observer's point of view. An object directly overhead has an elevation angle of 90°, complement of zenith distance.

Elevation mask (angle) An elevation angle used in a GPS receiver as the lower limit for using a satellite. Satellites that are closer to the horizon tend to have greater signal fade and experience greater multipath effects. Therefore, most GPS receivers have a minimum elevation angle and will not use the measurements from satellites whose elevation angle is below that limit. Also called horizon mask. Typical values in use range from 0 to 15°.

Ellipsoidal height Altitude as measured from the reference ellipsoid, as opposed to geoidal height (q.v.). GPS receivers generally compute altitude above the ellipsoid because positions are actually computed in ECEF coordinates (q.v.) and then converted to ellipsoid values using the definition of the reference ellipsoid. Conversion between ellipsoid and geoid height requires the use of a mathematical model.

Ephemeris The description of a satellite's orbit. As used in GPS, ephemeris refers to the data sent in subframes 1, 2, and 3 of the navigation message, and includes not only the orbit description but also the information needed to correct that satellite's clock to match GPS time. The orbital description consists of Keplerian elements which describe the shape, orientation, and location of the orbit with respect to the Earth, and refinements to the orbit called harmonic coefficients which permit more precise descriptions of the actual orbit. Plural of ephemeris is ephemerides (pronounced ef-em-*air*-i-deez).

Epoch An interval of time, as a single satellite measurement interval, or a specific point in time, as the time at which a position solution is valid. There is not one specific meaning of this term, as it is used differently by various authors.

Frame The collection of all subelements of the GPS satellite navigation message. A frame is composed of five subframes broadcast in sequence and requires 30 s to broadcast. It contains a total of 1,500 total bits of data and parity.

GDOP Geometric Dilution of Precision, a number which helps estimate total error in a GPS solution. GDOP is computed from the total number of satellites used in a solution, and their respective azimuths and elevations. As the total number of satellites increases, and their dispersion about the sky increases, GDOP decreases. GDOP can be broken down into PDOP and TDOP.

Geodetic survey A survey generally of higher precision and covering a greater area than a normal land survey. Such a survey would typically take into account a mathematical model of the Earth (an ellipsoid) and would be undertaken in order to establish references for subsequent land surveys or observations.

Geodesist A scientist who specializes in measuring the Earth.

Geoid A surface on the Earth of constant gravitation force equal to mean sea level (MSL). The geoid is an imaginary surface with significant undulations since the local gravitational force is affected by such things as mountains, the shape of the sea bottom, and other physical features.

Geoidal height Altitude above the geoid, also called altitude above MSL. This is the common altitude generally reported on maps and charts. See ellipsoidal height.

Gold code A type of PRN, specifically one of the family used to modulate the L1 carrier on the GPS satellites. Gold codes are created by combining the outputs of two linear feedback shift registers (LFSR) clocked at the same time. In GPS systems, the Gold codes are 1,023 bits long and have nearly the same number of 1s and 0s in each sequence. There are 37 Gold codes (of which two are identical, making 36 unique codes) possible using the specific LFSRs used in GPS satellites. To create the Gold codes, each LFSR is initialized to a starting point and then one is delayed a specific number of bits from the other before the outputs are combined.

GPS observation A set of measurements made on a single GPS satellite at a single time. Each observation can consist of multiple observables, including the pseudo-range as measured from the C/A code or P code, the carrier phase made on the L1 or L2 carrier, and estimates of the errors on each observable.

GPS time, GPS week The GPS system measures time in weeks and seconds into the current week. Each week starts at midnight between Saturday and Sunday, and consists of 604,800 s. The weeks are counted from Sunday, January 6, 1980, which was week 0. GPS time is computed by the GPS Control Segment, which averages all the atomic clocks in the system, including those on each satellite and those in the control stations. The Control Segment steers GPS time to align closely with UTC (typical difference between the start of a GPS second and the start of a UTC second is less than 20 ns). However, GPS time differs from UTC in that GPS does not add in leap seconds as UTC does from time to time. As of the end of 2002, GPS time was 13 s ahead of UTC due to the addition of that many leap seconds by UTC since 1980.

HDOP HDOP, a number that relates to the amount of error in a GPS solution in the local horizontal plane. HDOP is one component of PDOP (q.v.).

Height (ellipsoidal) The distance of a point above a reference ellipsoid surface. In GPS systems, the surface is usually the WGS84 reference ellipsoid.

Horizon mask See elevation mask.

Inertial Navigation System (INS) A navigation system that uses inertial instruments such as gyroscopes and accelerometers to measure changes in vehicle velocity, position, and attitude. Such a system generally requires a starting position, but thereafter computes the current position by measuring the changes. An INS is generally considered very accurate in the short term, but is subject to errors that accumulate over time from the last time the system was given a position update.

Integer bias The number of whole carrier phase cycles between a GPS receiver and satellite. Subject to great ambiguity, and determined by a process called ambiguity resolution.

Ionosphere The collection of ionized particles which form a layer usually between 50 and 1,000 km above the Earth. The ionosphere affects radio signals traveling through it, in an amount that varies inversely with the frequency of the signal. With spread-spectrum signals like GPS, the effect is to advance carrier phase and delay modulation codes. The amount that a signal is affected by the ionosphere varies in a daily cycle, and to a lesser extent, over the solar cycles. GPS satellites transmit two signals (L1, L2) so that receivers equipped to receive both signals can determine the effect of the ionosphere by comparing the changes with respect to frequency.

Keplerian elements Factors which together describe the orbit of one body about another. The elements consist of factors describing the shape of the ellipse (length of semi-major axis and eccentricity), the orientation of the ellipse with respect to the Earth (longitude of ascending node, inclination angle, and argument of perigee), and the location of the satellite in the orbit at some specified time (mean anomaly and mean motion difference). GPS satellites broadcast these elements, along with some additional terms to refine the description, in subframes 2 and 3 of the navigation message.

Kinematic Generally used in GPS to refer to a differential GPS mode where two receivers communicate over a real-time data link (real-time kinematic, RTK) and mutually resolve the integer ambiguity of the carrier phase. As a result, one receiver, operating in a fixed location, acts as a reference station, and the other, which can be mobile, is able to generate relative positions with respect to the reference station at sub-meter accuracy. Better systems can generate positions at the sub-centimeter accuracy.

L1, L2 The two carrier frequencies transmitted by the GPS satellites. L1 is transmitted at 1,575.42 MHz, while L2 is at 1,227.60 MHz. Both of these frequencies are in the microwave L-band, thus the names. Historically, the modulation on the two has differed. L1 carries the C/A code at 1.023 MHz and the P or Y code at 10.23-MHz chipping rates. L2 only carried the P or Y code at 10.23 MHz. However, there are plans to add the C/A code to the L2 carrier so that users who are not authorized to use the P code can still resolve ionospheric delays. Note that the use of spread-spectrum techniques, with each satellite using a unique pseudorandom number (PRN identifier), permits all GPS satellites to use the same L1 and L2 frequencies simultaneously, with receivers separating the signals using any of the several "despreading" techniques.

Model range The range between a satellite's broadcast position and a receiver's known position. The difference between model range and pseudo-range (q.v.) is the value generally computed and broadcast by differential GPS reference receivers.

Multipath A situation where a radio signal such as a GPS signal arrives at a receiving antenna from more than one direction. Typically, this is the result of the direct signal reaching the antenna and another copy of the same signal also arriving after reflecting off a reflective surface. This is a common GPS error source, since the presence of the reflected signal can confuse the receiver's tracking circuits and make the range measurements less precise.

Navigation message The data stream modulated onto the GPS PRN signal. It is added to the data stream at 50 bps by EXORing the data stream with the PRN code, resulting in some of the 1-ms PRN sequences being inverted. The message contains the satellite's ephemeris, almanac data, and other elements, organized into words, subframes, and frames (q.v.).

P code, Precise code The second, more precise signal modulated onto the L1 and L2 carriers. It consists of a PRN sequence modulated at a 10.23-MHz chipping rate (equivalent to a 29.3-m wavelength). The sequence is long enough that it does not repeat during an entire week. When the anti-spoofing feature is activated in a satellite, the P code is encrypted and is then referred to as a Y code.

PDOP Position dilution of precision, a measure of the effect that current satellite geometry has on the accuracy of position solutions. PDOP can be broken down to HDOP and VDOP, and can be combined with TDOP to create GDOP. See dilution of precision and the other DOP entries.

Phase See carrier phase.

Phase center, phase-center offset The point in space that corresponds to where the signal appears to be received by an antenna. Phase center is generally only of interest in very precise surveying applications, since the variation between L1 and L2 signals is typically on the order of millimeters. Phase-center offset is often computed using precise RF measurements and reported with respect to some easily accessible landmark on the antenna structure so that surveyors can align it with benchmarks or other survey points.

Point positioning The process by which observations made at a station are combined to compute the location of that point relative to the reference system. In GPS work, the GPS observations which determine a range between the satellite and the receiving antenna are combined with the position of the satellite at the time of the measurement, as computed from the ephemeris. This term relates to a single, independent solution in contrast with navigating (computing a continuous series of fixes, on what may be a moving platform).

Precise ephemeris Ephemeris for a satellite generally computed from external measurements (such as from several fixed locations, or from laser or optical observations) that can be used to define the satellite's orbit and location more precisely than the ephemeris given in the broadcast message. Precise ephemerides are computed by several groups (e.g., US National Oceanographic and Atmospheric Administration, and International GPS Service for Geodynamics) and may be posted on the Internet for use in high-precision post-processing.

PRN code, identifier PNR code or identifier. The identification of the specific pattern of 1s and 0s used by each GPS satellite to spread its carrier spectrum. For the C/A code, the patterns are called Gold codes, and there are 36 defined. Only 32 are actually available for use in the satellites since there are only 5 bits available in the navigation message to designate each satellite.

Pseudo-range The measured range between a satellite and a receiver using the "Time Received – Time Sent" equation. Its called "pseudo-range" because as it is used in the solution for user position, it has the receiver's clock bias added in. In other words, if the receiver clock bias was zero, the ranges would be nearly perfect or the "best estimate."

RAIM Receiver autonomous integrity monitoring (RAIM) – a software process inside a receiver that monitors time or position solutions in an attempt to detect and remove contributions by satellites that might be contributing to excessive errors. This process generally requires the availability of more satellites than are required for a solution so that aberrant contributors can be detected.

Rapid static A receiver point-positioning mode similar to static, but with modifications to permit faster solutions. Typically, this involves larger error limits than static mode.

RTK A differential GPS method that involves tracking and resolving ambiguity on the carrier phase. A reference receiver provides dynamic tracking information from a reference location, and the remote receivers perform double differencing to remove satellite and receiver clock errors. Resulting position solutions can be computed to sub-meter or sub-centimeter precision in many cases.

Reference station, reference receiver A GPS receiver used to compare with other receivers in order to improve position solution precision. The reference receiver is typically located at a fixed and known position, and may either provide differential GPS corrections, or collect data for use in post-processing the data from the other receivers.

Right-hand coordinate system A coordinate system where the X, Y, and Z axes can be represented, respectively, by the thumb, index, and middle fingers of the right hand, each arrayed so as to be mutually perpendicular. The WGS-84 system uses such a coordinate system.

RINEX Receiver INdependent EXchange format, a file format defined for standardizing data exchange between GPS receivers. This format has been adopted by the International Association of Geodesy (IAG) to permit data exchange between different receiver brands.

Sampling interval The period of time during which GPS satellite observations are collected and combined to form a single output. Receivers may sample for periods from 20 ms to several seconds, depending on the particular receiver design and application.

Selective availability (SA) An intentional error added to the satellite's C/A code under certain circumstances. SA limits the precision of position solutions by adding artificial errors to the clock, the reported ephemeris, or clock corrections of the satellites. SA can be added to the signals at various levels to induce position errors of up to 150 m. Do not confuse SA with A-S (q.v.).

Single difference A GPS processing method that combines observations of pseudo-range or carrier phase between two receivers, each observing the same satellite at the same time. When the observation from one receiver is subtracted from the observation of the other receiver, the resulting single difference value is free of any clock error in the satellite, any satellite frequency offset has been reduced, and satellite hardware delay is eliminated. The resulting value can be further used to create double or triple differences (q.v.).

Space segment One of the three elements of the GPS system. The space segment consists of all the satellites in orbit. The complete GPS design calls for a total of 24 satellites, with four satellites spaced evenly in six separate orbits. Each orbit is inclined about 55° to the equator and is spaced at approximately 60° from adjacent orbits.

Spread spectrum A radio signal modulation technique characterized by an output that has a very wide bandwidth and low average amplitude over that bandwidth. It is created by altering the phase of the carrier at an apparently random rate called the chipping rate. In GPS, the carrier is 1,575.42 MHz for the L1 carrier, and the modulating signal is a 1.023-MHz stream of pseudo-random 1s and 0s. Whenever the bit changes, the carrier's phase is inverted. The resulting signal is spread across the spectrum in a pattern which looks like the function $\sin^2 x / x^2$ with the central peak 2.046-MHz wide, and the side peaks 1.023-MHz wide.

Static A receiver operating mode where the receiver is left stationary over the point to be surveyed, and multiple observations are collected. Such a survey can last from a few seconds to hours or even days. Also called fixed.

Station Surveyor's term for a location to be surveyed.

Stop and go A survey method where two GPS receivers are used. One is established at a fixed site, while the other is first allowed to acquire satellites, then it is moved from station to station, where it is positioned momentarily, some form of marker is attached to the data, and then the receiver is moved to the next location. During the entire time, the two receivers maintain lock on the satellites so that ambiguity resolution is maintained.

Subframe A unit of data in the GPS navigation message. Each subframe is composed of ten data words, each of which are 30 bits long. Thus, the subframe is 300 bits long and requires 6 s to transmit at 50 bits per second. There are five defined subframes, which when broadcast together in sequence comprise a frame.

TDOP Time dilution of precision, a measure of how much error exists in the time solution. Unlike other DOPs, TDOP does not vary with location of the satellites, but only with the number of satellites used. TDOP and PDOP together comprise the orthogonal components of GDOP. See DOP, GDOP, and PDOP.

Triple difference A technique of processing GPS observations that involves computing the difference between two double differences (q.v.) taken using the same two satellites and two receivers at two different times, usually taken one observation interval apart. Triple differences are free of any integer ambiguity found in the double differences and can be used to detect cycle slips in the carrier observations.

Troposphere The layer of the Earth's atmosphere located closest to the surface of the Earth. Its thickness varies from over 16 km at the equator to less than 9 km at the poles. It is the layer of the atmosphere where most of the weather occurs.

UERE User Equivalent Range Error (UERE), an estimate of the errors in a pseudo-range measurement from system components such as the satellite clock and orbit, ionospheric and tropospheric effects, and hardware delays. Satellites broadcast an estimate of the UERE.

User Segment The collective name for all receivers listening to GPS satellites. It is one of the three defined segments (see Control Segment and Space Segment), but it has no specific organization.

UTC Coordinated universal time, the uniform atomic time defined internationally and maintained in the USA by the United States Naval Observatory; it is used as the basis of GPS time. GPS time differs from UTC by leap seconds, which are periodically added to UTC but not to GPS time, and by some small fraction of a second which is generally kept under 20 ns.

VDOP Vertical dilution of precision, the vertical component of system errors due to satellite geometry. It is most influenced by the dispersion of satellites in elevation, which is limited by the horizon. As a result, VDOP will normally be larger than HDOP from a given satellite constellation. VDOP and HDOP together form the orthogonal components of PDOP. See DOP, HDOP, and PDOP.

WGS84 Datum, Ellipsoid World Geodetic System 1984, the standard datum used by the GPS system, and its associated ellipsoid. The semi-major axis (distance from the center of the Earth to the equator) of the ellipsoid is set at 6,378,137.0 m, and the flattening value (flattening is defined as the difference between the semi-major and the semi-minor axes divided by the length of the semi-major axis) is 1/298.257223563. On this ellipsoid, an ECEF coordinate system is defined, with the X and Y axes defining the equatorial plane, and the Z axis aligned with the Earth's spin axis.

Word An element of the navigation message consisting of 30 bits. The first 24 bits actually contain data, while the last 6 bits are data-integrity bits (usually called parity bits, but actually 6 bits of an 8-bit Hamming code). A collection of ten words together form a subframe (q.v.).

Y Code The encrypted form of the P code broadcast when A-S is active (see A-S and P code).

Index